***FM 4-30.31 (FM 9-43-2)/MCRP 4-11.4A (FMFRP 4-34)**

Field Manual
No. 4-30.31
Marine Corps Reference Publication
No. 4-11.4A

Headquarters
Department of the Army
United States Marine Corps
Washington, DC, 19 September 2006

Recovery and Battle Damage Assessment and Repair

Contents

***This publication supersedes FM 9-43-2/FMFRP 4-34/TO 36-1-181, 3 October 1995.**

Figures

Tables

Preface

This manual provides the authoritative doctrine guidance on using recovery and repair assets on the battlefield. Practical methods of recovering or repairing equipment (disabled or immobilized) due to hazardous terrain, mechanical failure, or a hostile action are also addressed.

Field manual (FM) 4-30.31 is directed toward both the leader and the technician. Tactically, it provides an overview of how recovery and battle damage assessment and repair (BDAR) assets are employed on the battlefield. Technically, it provides principles of resistance and mechanical applications to overcome them. Equipment, rigging techniques, and expedient repairs are summarized as a refresher for recovery-trained military personnel and as general guidance for others.

The procedures and doctrine in this manual apply to both wartime operations and military operations other than war. Normally, BDAR should be used when and where standard maintenance practices are not practical because of the mission, enemy, terrain and weather, troops and support available, time available, civil considerations (METT-TC) or METT-T space and logistics (METT-TSL) for USMC. BDAR is not intended to replace standard maintenance practices but rather to supplement them under certain conditions. Standard maintenance procedures provide the best, most effective means of returning disabled equipment to the operational commander—provided adequate time, parts, and tools are available.

High-risk battle damage repairs (involving possible danger to personnel or further damage to equipment) are only permitted in emergencies, normally in a battlefield environment, and only when authorized by the unit commander or his designated representative. The goal is to return a combat system to the battlefield in the least amount of time, while minimizing danger to personnel and equipment.

BDAR techniques are not limited to simply restoring minimal functional combat capability. If full mission capability can be restored expediently with a limited expenditure of time and assets, it should be restored. This decision is based on METT-TC.

Some BDAR techniques, if applied, may result in shortened lifespan or further damage to components. The commander must decide whether the risk of having one less piece of equipment outweighs the risk of applying a potentially destructive field-expedient repair. Each technique provides appropriate warnings and cautions, which list the system's limitations caused by the action. Personnel must use ground guides and extreme caution when operating recovery assets around or on an aircraft.

The proponent of this publication is the United States Training and Doctrine Command. Users of this manual are encouraged to submit suggestions, changes, or comments to improve this manual. Comments with justifications should be keyed to the specific page, paragraph, and line of text; prepared on Department of the Army (DA), DA Form 2028 (Recommended Changes to Publications or Blank Forms); and forwarded to Commander, U.S. Army Combined Arms Support Command, ATTN: ATCL-AO, Fort Lee, VA 23801-6000.

The provisions of this publication are subject to international Quadripartite Standardization Agreement 171.

This publication applies to the United States Marine Corps, Active Army, the Army National Guard/Army National Guard of the United States, and the U.S. Army Reserve unless otherwise stated.

This is a joint service manual. Other Armed Forces should use their appropriate service forms in place of the DA forms listed herein. All Department of Defense (DD) forms listed in this manual will be used by all Armed Forces.

Basic issue items (BII) and additional authorization list (AAL) are Army terms. The Marine Corps' equivalent of these terms is on-vehicle equipment (OVE).

Chapter 1

Introduction to Recovery and Battle Damage Assessment and Repair

Recovery and battle damage assessment and repair (BDAR) are subsets of maintenance. Both are the owning units' responsibilities; both have a fundamental purpose of returning combat assets to the battlefield as soon as possible. The purpose of recovery is to rapidly remove disabled or mired equipment from the battlefield, while the purpose of BDAR is to rapidly repair the equipment to continue the mission or allow self-recovery. Recovery vehicles, both wheeled and tracked, should carry a minimum of one BDAR kit to assist in recovery operations.

RECOVERY

1-1. Recovery is retrieving, that is, freeing immobile, inoperative, or abandoned equipment from its current position and returning it to operation or to a maintenance site for repair. These actions typically involve towing, lifting, or winching. Towing is usually limited to moving equipment to the nearest unit maintenance collection point (UMCP). Recovery consists of—

- Self-recovery: Actions require using only the equipment's assets.
- Like-recovery: Actions involve assistance from a second, like, or heavier class vehicle.
- Dedicated-recovery: Actions require assistance from a vehicle specifically designed and dedicated to recovery operations.

1-2. Unless specifically mentioned, recovery tactics, techniques, and procedures, and doctrine, organization, training, materiel, leadership and education, personnel, and facilities considerations apply to both combat and noncombat range of military operations (ROMO).

SELF-RECOVERY

1-3. Self-recovery starts at the location where the equipment becomes mired or disabled. The operator/crew uses the BII and AAL or OVE items to perform self-recovery.

1-4. When the equipment has a mechanical failure, the operator/crew will use the equipment's technical manual (TM) to perform troubleshooting procedures with the tools available in the BII and AAL or OVE. When self-recovery fails, the operator/crew can request assistance from available like vehicles.

Note. According to current Army doctrine, an equipment self-recovery winch can be used only to recover the equipment on which it is mounted. Self-recovery winches should not be used to recover other mired equipment. Self-recovery winches can be used to assist in a recovery effort by providing stabilizing or holdback capabilities. Refer to the equipment operator's manual for like-vehicle recovery procedures and limitations. This does not apply to the United States Marine Corps (USMC).

LIKE-RECOVERY

1-5. Like-vehicle recovery is used when self-vehicle recovery fails. The principle is to use another piece of equipment—of the same weight class or heavier—to extract or tow the mired equipment by using tow bars, chains, tow cables, and/or allied kinetic energy recovery rope (AKERR). When self-recovery and like-recovery are not practical or are unavailable, use dedicated recovery assets.

Note. AKERR is used to extract mired equipment; it is not designed as a towing device (see chapter 4 for more details on AKERR).

Dedicated Recovery

1-6. Dedicated-recovery vehicles are used when self-recovery or like-vehicle recovery is not possible because of the severity of the situation, safety considerations, or the inability to use like-vehicle assets employed in their primary mission. Recovery managers and supervisors must ensure recovery vehicles are used only when absolutely necessary. Dedicated recovery vehicles must be returned as quickly as possible to a central location to support the unit. In addition to its recovery mission, this equipment is often used for the heavy lifting required in maintenance operations. Recovery managers and supervisors must use all available resources carefully to provide sustained support.

BATTLE DAMAGE ASSESSMENT AND REPAIR

1-7. BDAR is the procedure used to rapidly return disabled equipment to the operational commander by field-expedient repair of components. BDAR restores the minimum essential combat capabilities necessary to support a specific combat mission or to enable the equipment to self-recover. BDAR is accomplished by bypassing components or safety devices, cannibalizing parts from like or lower priority equipment, fabricating repair parts, jury-rigging, taking shortcuts to standard maintenance, and using substitute fluids, materials or components. Depending on the repairs required and the amount of time available, repairs may or may not return the vehicle to a fully mission-capable status. Operators/crew, maintenance teams (MTs), maintenance support teams (MSTs), combat repair teams (CRTs), or recovery teams may perform BDAR.

Peacetime Battle Damage Assessment and Repair Applications and Training

1-8. Army Regulation (AR) 750-1, Army Maintenance Management Policy, paragraph 8-8, and USMC's TM-4700-15/1 require unit commanders to conduct nondestructive peacetime BDAR training. Soldiers must become familiar with the components in the BDAR kits that enable many repairs which otherwise would not be possible. Each crewmember should know how to perform battle damage assessment (BDA) for assigned equipment.

1-9. Maintenance standards are based on TMs and preventive maintenance checks and services. Low-risk, nondestructive BDAR can be applied as training during a ROMO. The goal of training is to simulate combat conditions as closely as possible. Commanders can direct the employment of BDAR for normal maintenance failures to evaluate training or validate new procedures. Unit commanders should develop sustainment training in which vehicle crews and field maintenance mechanics conduct BDAR and recovery operations. Skills required to perform BDAR are found in individual and collective training tasks. Unit commanders should know which military occupational specialties require knowledge of BDAR. Company and battalion Army Training and Evaluation Program tasks, especially in combat service support units, require some knowledge of BDAR. Military qualification standards also require ordnance officers of all grade levels to have awareness of BDAR policy and doctrine incorporated in developmental training.

1-10. Peacetime BDAR training is necessary for Soldiers to perform these vital tasks on the battlefield. Recent military operations have highlighted how U.S. forces can be deployed almost overnight into situations where combat is expected and normal logistical arrangements are underdeveloped or unavailable.

1-11. Mechanics should master BDAR techniques very quickly since they have already received maintenance and mechanical training. Vehicle crewmembers should be familiar with BDAR techniques for their specific equipment. Cross training of mechanics is also necessary. Crewmembers and mechanics that can change a fan belt or fuel filter can easily learn to perform BDAR for these same items. Experience in live-fire tests at Aberdeen Proving Ground, MD, and Meppen, Germany, has shown that Soldiers can learn BDAR procedures with minimal training. With the adoption of the U.S. Army Tank-automotive and Armaments Command ground equipment BDAR kits and easy-to-use instructions, the commander's task of training will be much simpler. In addition, the BDAR training techniques outlined in BDAR TMs are straightforward tasks with easy-to-comprehend illustrations.

1-12. Training BDAR does not require expensive or complicated training aids. During field exercises, low-risk BDAR repairs can be applied to a unit's equipment until standard maintenance procedures can be performed, if required. Service members must use BDAR TMs and kits to become proficient in their use and assessment procedures. If more sophisticated or difficult tasks are to be trained, unserviceable radiators, fuel tanks, hydraulic lines, and condition code "H" equipment may be acquired through the local Defense Reutilization and Marketing Office (DRMO) to support such training. Supervisors at all levels should use the low-, medium-, and high-risk classifications when performing such training to minimize accidents and/or damage to equipment. The unit commander decides whether to use low-risk peacetime BDAR during training.

1-13. To summarize, a unit BDAR training program should incorporate the following key training parameters:

- Orient crews and maintainers on current BDAR doctrine and policy.
- Train personnel in using BDAR TMs.
- Train both crews and maintainers in using the crew and maintainer BDAR kits.
- Periodically review BDAR training films.
- Train personnel to identify BDAR repairs that can restore equipment to a mission-capable status after a breakdown. This restoration includes the following:
  - Installing components from other equipment that can be modified to fit or interchange with components to restore basic functional capability.
  - Fabricating parts, performing field-expedient repairs, and using substitute fuels, fluids, and lubricants.
  - Bypassing or shortcutting components to restore operation of damaged mission essential elements.
  - Allocating time in the unit-training schedule to practice BDAR and recovery.
  - Using local materials from the DRMO or other available equipment (such as vehicle hulls used as range targets) for hull patching. This will be helpful in familiarizing personnel with BDAR kits.

Chemical, Biological, Radiological, and Nuclear Environment

1-14. BDAR and recovery may be required in a chemically contaminated area or under adverse conditions with severe limitations involving personnel, facilities, equipment, and materials. It may be necessary to perform BDAR and recovery tasks while wearing protective gear. If the situation does not allow a recovery team to recover the equipment, personnel should keep BDAR to a minimum, and then recover to a decontamination area. Personnel can finish the BDAR or recovery task when decontamination procedures have been completed.

Note. For decontamination procedures, refer to FM 3-5 / Marine Corps Warfighting Publication (MCWP) 3-37.3.

1-15. Recovery teams must be trained in chemical, biological, radiological, and nuclear (CBRN) defense procedures, monitoring, and detection techniques. The teams should have additional decontamination equipment, decontaminating agents, and protective clothing. Contaminated recovery equipment could spread an agent along the evacuation route and pose a hazard to uncontaminated units and equipment. Commanders should keep in mind that equipment used to recover contaminated equipment will also be contaminated. Contaminated equipment should not be evacuated to uncontaminated areas until it is fully decontaminated. The sooner the contamination is removed, the sooner mission-oriented protective posture (MOPP) levels can be reduced, and restoration of combat power can begin.

Depleted Uranium

1-16. A complete assessment must be performed before entering or attempting to repair any damage. Damaged ammunition that has depleted uranium (DU) as a component material includes the risk of radioactivity.

1-17. DU is a dense, slightly radioactive metal used by the United States and other forces in munitions, armor, and other applications. DU has properties similar to the more familiar heavy metals—such as lead and tungsten. Under certain conditions, Soldiers could inhale or ingest DU "dust" or sustain injuries that could result in DU intake exceeding U.S. safety standards. Following the precautions outlined in this section will help protect personnel against DU and other heavy metal contaminants.

1-18. Personnel could receive a dose (or intake) of DU that exceeds U.S. safety standards if personnel are in, on, or near an armored combat vehicle when it is struck by DU munitions, or when an M1A1/A2 heavy armor tank is struck and breached by any munitions (DU or non-DU). When a DU round hits and penetrates armor (specifically DU armor), fragments, oxides, and other particles are formed. DU can be taken into the body through inhalation, ingestion, or wound contamination or as embedded fragments. (DU particles settle soon after a munitions strike, reducing the risk of DU intake by inhalation.)

Depleted Uranium Preventative Measures

1-19. If any munitions penetrate the crew compartment of an armored vehicle, eventually the inside of the vehicle will have to be decontaminated to remove all possible hazards. Such hazards could include DU contamination, as well as contamination from burned plastics; petroleum, oils, and lubricants (POL) products; and tungsten. When mission, enemy, terrain and weather, troops and support available-time available (METT-T) permit, decontaminate the crew compartment of the vehicle according to procedures in FM 3-5/MCWP 3-37.3. While decontaminating the crew compartment of a damaged vehicle, DU contamination could be re-suspended, as well as contamination from burned plastics, POL products, and tungsten that had previously settled inside. While performing these actions, some type of respiratory protection (such as a dust mask, protective mask, or other approved respirator) should be worn. Personnel should cover exposed skin. (An increase in MOPP is not required.) After leaving the vehicle and before removing the protective mask, personnel should dust off the uniform worn while performing the decontamination operations. They must always observe standard field hygiene procedures, to include washing their hands and face.

1-20. Personnel must remain at least 50 meters from actively burning equipment involving DU. Personnel should also treat all wounds involving DU in the same manner as any other wound and then notify medical personnel. Additional training can be obtained by consulting the U.S. Chemical School Depleted Uranium Training at: http://www.wood.army.mil/84chem/HHC/TTD/Rad%20Lab/du.htm

Environmental Protection

1-21. Fuel, oil, and other mechanical fluids spilled on the ground during BDAR operations can cause damage to the environment. As with many BDAR considerations, the level of environmental protection will be mission-dependent. During periods of heightened conflict, simple procedures can help to preserve and protect our fragile environment. All practical efforts should be made to avoid environmental contamination. Spills of more than one gallon should be reported through the chain of command to the unit's logistical element, such as the battalion S-4. Local policy or state laws may require spills of even lesser amounts to be reported.

Tools and Equipment

1-22. The Training and Doctrine Command (TRADOC) Executive Agency for BDAR and Recovery has designed special BDAR kits for ground equipment. These kits are use at the breakdown site to give the operator/crew or maintenance personnel the capability to perform BDAR without access to special tools, parts, and equipment. Maintainer kits are normally kept with the maintenance team, CRT, or the MST. A smaller, lighter, less expensive crew-level kit is for the operator/crew to use aboard vehicles. BDAR is not limited to what can be performed with these kits, but they do provide the crew and maintenance personnel with materiel and tools on hand specifically designed for BDAR. Vehicle BII, organizational tools, and tools found with the contact teams are to be used with the BDAR kits.

1-23. Each unit is expected and encouraged to modify its kits to suit its special operational needs and geographical environment. (See appendix A for additional information on BDAR kits.)

> ***Note.*** Host-nation support (HNS) may be used to offset problems encountered by units deployed overseas.

1-24. Units deployed overseas find themselves at the end of a long line of communications. In addition, Soldiers operate in harsh terrain and less than desirable climatic conditions. Units may expend their basic load of supplies quickly and may face a long wait for replenishment. Deployment flow may not allow delivery of recovery assets until late in the deployment sequence. To offset these problems, HNS can be used to replenish contents of BDAR kits and provide recovery assets, materiel, and equipment. Even in the most remote locations of the world, HNS will usually have industrial resources and automotive repair capabilities and materiel suitable for recovery. Procurement officers should be able to replenish BDAR kits and obtain other supplies and tools required to perform BDAR and/or recovery operations. Higher headquarters publishes command directives outlining HNS procurement procedures and sources of supply. BDAR kit contents are selected and designed for replenishment from local sources. Commanders and maintenance personnel must be prepared to avail themselves of this valuable resource.

1-25. Host or hostile nation service stations, industrial facilities, and automotive service and supply centers are invaluable sources of supply. The responsible field procurement officer must closely supervise the use of local sources. Prompt payment of purchases and payment for rental space and facilities is crucial to maintaining friendly relations with the local population and retaining its support. If using abandoned HN or hostile nation equipment suitable for BDAR and recovery operations, personnel should gain command approval prior to use. Procurement offices should be involved in the HNS decision.

1-26. During operations in a HN, such as ROMO missions, U.S. forces will comply with all applicable regulations, including status-of-forces agreements, treaties, and international agreements. HN laws governing spills, such as fuel and oil, are important to BDAR. If HN laws do not exist, U.S. forces will follow the laws of the United States. (See appendix B for information on combined and multinational BDAR operations.)

Chapter 2

Principles of Recovery

Recovery assets are normally centrally managed. This chapter provides direction, permits better management, and promotes quicker responses to task organization, workload, and the tactical situation. In combat units, the recovery manager is designated at battalion level and is normally the battalion maintenance officer (BMO) [Army] or battalion motor transport officer (MTO) [USMC]. Some combat arms battalions (for example, nondivisional, defense battalions, engineer battalions, and so forth.) have company-sized units that operate without a BMO or battalion maintenance team. In these types of units, the senior maintenance supervisor manages these assets.

GENERAL

2-1. Recovery is performed to—

- Retrieve damaged or mired equipment for repair and return to use.
- Retrieve abandoned equipment for further use.
- Prevent enemy capture of equipment.

2-2. The type and quantity of supported equipment, as well as the tactical situation, may require tailoring of recovery assets. Only the minimum number of required recovery assets should be deployed for each mission. The BMO or other designated individual coordinates recovery operations with overall repair efforts to support the commander's priorities and tactical situation most effectively. The following general principles apply to recovery management:

- Using units are responsible for their own equipment. The crew of using units should first attempt self-recovery efforts, followed by unit-supported like-vehicle recovery. If necessary, limited backup support is available from the next higher level of maintenance.
- Using units are responsible for security during the entire recovery mission.
- Recovery operations are coordinated with the maintenance effort.
- Recovery vehicles of the correct weight class must be used to ensure safety. Recovery vehicle winches and/or towing capabilities must be greater than those of the disabled or mired vehicle.
- Recovery vehicles should not return equipment farther than the UMCP. This keeps recovery assets forward.
- Recovery teams must use CBRN contamination avoidance principles to avoid contamination and minimize targeting.
- Recovery teams must take all practical steps to avoid spills and other environmental contamination.

FUNDAMENTAL METHODS OF RECOVERY

2-3. A Soldier must answer the following questions when attempting a recovery task:

- What must be done?
- Which equipment must be used?
- Which techniques must be used?

WINCHING, LIFTING, TOWING, AND EXPEDIENTS

2-4. This chapter will answer these questions, beginning with a summary of the four methods of recovery. It will provide details about recovery tackle and how to use and maintain it. The four methods of recovery include—

- Winching, using winches on special purpose or cargo vehicles. (According to current Army doctrine, an equipment self-recovery winch can be used only to recover the equipment on which it is mounted. Self-recovery winches cannot be used to recover other mired equipment.)
- Lifting, using the lifting capabilities of special purpose vehicles.
- Towing, using the towing capabilities of similar or special purpose vehicles.
- Expedients, used when other methods are not adaptable to the situation or when appropriate like vehicles or dedicated recovery vehicles are not available.

RECOVERY SAFETY

Note. Heed all safety warnings found in the appropriate operator's manuals for all vehicles performing recovery operations in order to prevent injury or damage to equipment. Ground chocks and spades have their limitations. If overloaded, the recovery vehicle can slide out of control. Excessive speed for conditions is a major contributor for towing accidents. Refer to the equipment operator's manual for safe towing speeds for various terrain conditions. Place ***safety first*** during recovery operations.

2-5. Recovery can be inherently dangerous unless safety is continually practiced. Each of the recovery functions (winching, lifting, towing, and expedients) must be performed with safety as the primary concern. Always follow the safety warnings found within this manual and in the operator's manual for both the recovery vehicle and the recovered vehicle or equipment. Maintaining an awareness of the following key factors and actions can help prevent damage to equipment and injury to personnel.

- Be aware that winch cables can break and backlash into personnel.
- Exercise extreme caution when towing.
- Check the operator's manual for guidance and to validate repair procedures.
- Establish minimum safe distances and clear all unnecessary personnel from the recovery site.

2-6. Know recovery equipment capabilities and limitations. Winches have tremendous power, and, if not properly secured to the disabled vehicle, winches can rip off tow lugs, bumpers, and other attachments that often become projectiles, injuring personnel and/or damaging equipment. Always follow the safe rigging guidelines in this manual. Keep all but the minimum required personnel away from the recovery area. Each recovery crewmember must know where other crewmembers are located at all times.

2-7. Towed tracked vehicles and some wheeled vehicles will not have any braking effect. The recovery vehicle must provide braking for the towed vehicle as well as itself.

Note. Some tracked vehicles may also require a holdback vehicle during towing operations.

2-8. Wrecker lift-towing operations also require extreme caution. The towed vehicle performs abnormally because the vehicle weight is not distributed on all wheels, and the wrecker steering control is degraded because of reduced weight on the front wheels.

2-9. Other recovery lifting actions also require extreme caution to prevent injury to personnel and/or damage to equipment. Suspended loads can drop or slide. If the crane has a remote control, use it to keep away from the action. Using the remote control will allow the operator to observe equipment movement and recovery actions and to locate other assisting crewmembers that are not visible from the fixed control station. Never exceed the limitations of the crane or its outriggers.

Recovery Equipment

Wire Rope

2-10. Refer to FM 5-125 for detailed information about wire ropes. This FM provides procedures for the care, use, and maintenance of wire ropes.

> **CAUTION**
>
> Always wear leather gloves when handling wire rope. Small frays in wire strands can cause severe lacerations to hands. Never slide the wire rope through hands, even when wearing leather gloves. Use the hand-over-hand method when inspecting or handling wire ropes. Kinked, frayed, or unlayed wire ropes are unserviceable and will not be used.

Blocks

2-11. Blocks consist of a shell or frame with one or more grooved wheels called sheaves (figure 2-1). The military uses two basic constructions—snatch and conventional blocks.

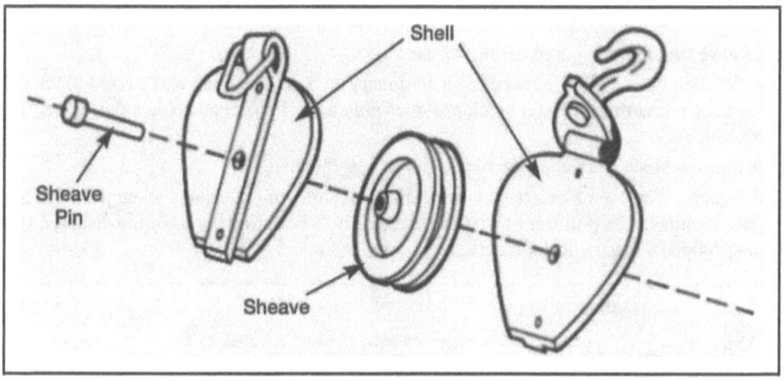

Figure 2-1. Block components

Snatch Block

2-12. A snatch block (figure 2-2) is used when the block will not be a permanent part of a tackle system (ropes, blocks, and pulleys used to raise and lower loads and/or apply tension); it can also be used as required based on the situation. A snatch block is constructed so the shell can be opened to admit a cable without reeving. Winch cables have attachment-like hooks or sockets on their free ends and can be reeved through a block.

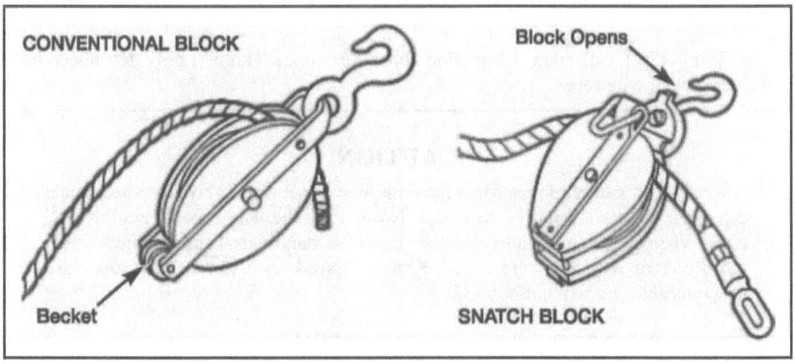

Figure 2-2. Block configurations

Conventional Block

2-13. A conventional block (figure 2-2) is generally used where it will remain as part of a rigging system. On recovery equipment, it is used with rope. To form a tackle with conventional blocks, lay out the blocks, and thread or reeve the wire rope through the blocks.

2-14. Blocks have the following applications (figure 2-3):

- A fixed block is a block attached to a stationary anchor. The sheave of a fixed block permits the rope to change direction. A fixed block provides only a mechanical advantage during self-recovery operations.
- A running block is an attached block that moves the load.
- A floating block is a block used with a tow cable, allowing the cable when pulled to align with the power source. The pull can be distributed equally to both tow hooks of the disabled vehicle and does not produce a mechanical advantage.

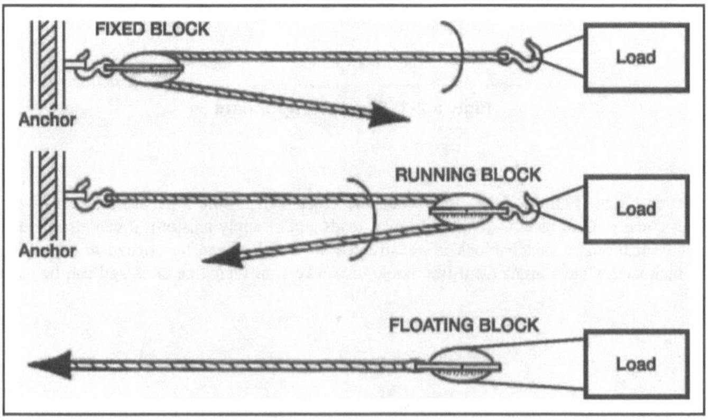

Figure 2-3. Block classifications

Chains

2-15. Refer to FM 5-125, Rigging Techniques, Procedures, and Applications for detailed information about chains. This FM also provides procedures for the care, use, and maintenance of chains.

Hooks

2-16. Refer to FM 5-125 for detailed information about hooks. This FM also provides procedures for the care, use, and maintenance of hooks.

Allied Kinetic Energy Recovery Rope

2-17. The AKERR is a multi-strand, woven, nylon rope used for like-vehicle recovery. The rope is connected between the mired vehicle and the towing vehicle. The towing vehicle accelerates, stretching the rope, which creates potential energy. When the rope is fully stretched, it transfers the energy to the mired vehicle, giving it a strong, sudden pull. Refer to TM 9-4030-200-10 for use and care of the AKERR.

RESISTANCE

2-18. Resistance is defined as opposition to movement. In recovery operations, resistance is caused most often by terrain features; such as mud, sand, water, or the recovery tackle itself. This section will focus on vehicles disabled by terrain conditions.

2-19. Two factors that can be applied during recovery operations to help reduce resistance are direction of travel of recovery and power applied to tracks. (Applied reduction factors discussed in the following paragraphs do not apply to wheeled vehicles.) Once load resistance is determined, apply effort to affect recovery.

Types of Resistance

2-20. Five types of resistance may occur when recovering vehicles are disabled by terrain conditions. They are grade, overturning, mire, water, and tackle.

Grade Resistance

2-21. Grade resistance occurs when a vehicle moves up a slope (figure 2-4). Grade resistance (including nosed-in vehicles) is estimated as equal to the weight of the vehicle plus cargo. Even though actual grade resistance may be less than the weight of the vehicle, the most resistance encountered on a grade is the weight of the disabled vehicle plus cargo.

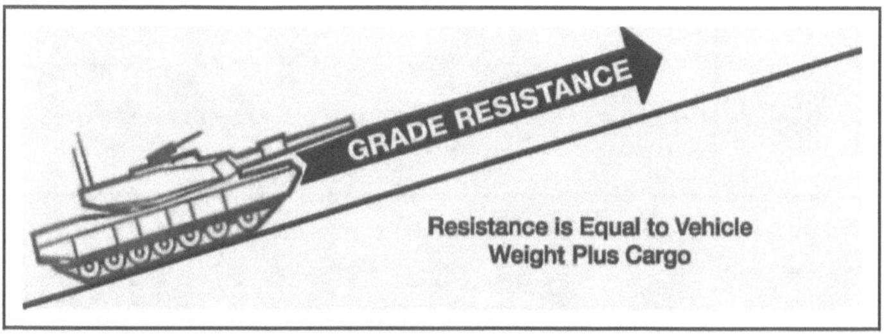

Figure 2-4. Grade resistance

Overturning Resistance

2-22. Overturning resistance is that weight of the vehicle that acts against the force exerted to bring it back on its wheels or tracks (figure 2-5). This force is approximately one-half of the vehicle's weight.

Figure 2-5. Overturning resistance

Mire Resistance

2-23. Mire resistance is created when mud, snow, or sand becomes impacted around the wheels, tracks, axle, gear housing, or hull of the vehicle. Mire resistance is described as wheel/track, fender, or turret/cab depth (figure 2-6).

2-24. Wheel depth mires occur when wheeled vehicles are mired up to the hub but not over the center. Tracked vehicles are mired up to the road wheels but not over the top. Estimate wheel-depth resistance as equal to the weight of the vehicle plus cargo.

2-25. Fender depth mires occur when wheeled vehicles are mired over the top of the hub but not over the fender. Tracked vehicles are mired over the top of the road wheels but not over the fender. Estimate fender depth mire resistance as twice the total weight of the vehicle plus cargo.

2-26. Turret or cab depth mires occur when vehicles are mired over the top of the fender. Estimate turret/cab depth mire resistance as three times the total vehicle weight plus cargo.

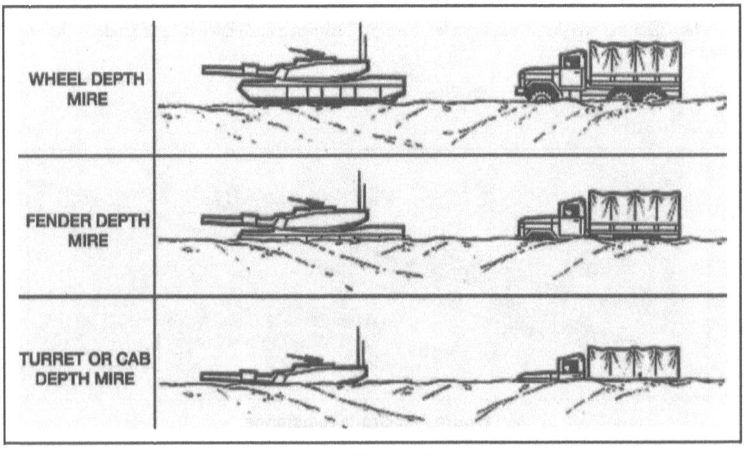

Figure 2-6. Mire resistance

Water Resistance

2-27. Water resistance occurs when submerged vehicles are pulled from water to land. Estimate the amount of resistance met in the same way as for land recovery. In some instances, the resistance to overcome is less than the rolling resistance of the same vehicle on land. (See chapter 4, paragraph 4-112, for more information.)

Tackle Resistance

2-28. Tackle resistance is that part of total resistance added to the recovery by friction in tackle. Tackle resistance is friction created by a sheave rotating in its pin, the rope flexing around the sheave, or the rope scuffing in the groove of the sheave, causing a loss in energy as the rope passes around the sheave. This loss is resistance and must be overcome before the load resistance can be overcome. Each sheave in the rigging will create resistance. To determine tackle resistance, multiply 10 percent (.10) of the load resistance by the number of sheaves (not blocks) in the rigging. For example, in the following calculations, the load resistance is 40 tons (80,000 pounds) and two sheaves are used.

Load resistance (vehicle + cargo)	= 80,000 lb
Number of sheaves	= 2
Tackle resistance	=.10 x 80,000 lb x 2 = 16,000 lb

CAUTION

Friction in tackle causes a loss in energy that must be overcome before the load resistance can be moved.

TOTAL LOAD RESISTANCE

2-29. Because tackle resistance must be overcome before the load resistance can be moved, the load and tackle resistance are added. This resistance is referred to as total resistance (the total amount of resistance the available effort must overcome). For example, in the previous example of tackle resistance, the load resistance of 80,000 pounds plus the tackle resistance of 16,000 pounds equals a total resistance of 96,000 pounds.

Load resistance (vehicle + cargo)	= 80,000 lb
Tackle resistance	= <u>16,000</u> lb
Total load resistance	= 96,000 lb

RESISTANCE REDUCING FACTORS

2-30. Situation and mechanical resistance affect the load resistance of mired vehicles. Resistance reducing factors do not apply to wheeled vehicles; they are only to be used for tracked vehicles.

Direction of Travel for Recovery

2-31. When a mired vehicle is recovered in the opposite direction of its travel, the tracks pass through ruts made by the vehicle when going into the mire. This reduces estimated resistance approximately 10 percent and is the preferred method of recovery. For example, a tank weighing 106,000 pounds is mired at wheel depth and can be recovered in the opposite direction of travel. Estimate resistance as 106,000 pounds and subtract 10 percent for recovery in the opposite direction of travel. The load resistance equals 95,400 pounds.

Vehicle weight	106,000 lb
Reduction factor	x .10
Estimated reduction	10,600 lb
	106,000 lb
	- 10,600 lb
Estimated load resistance	95,400 lb

Power Applied to Tracks

2-32. When power is applied to the tracks of a mired vehicle, the movement of the tracks helps to break the suction of mud against the belly of the vehicle. This reduces estimated resistance by approximately 40 percent. Before computing the 40 percent reduction, make sure the mire is not deep enough to prevent the operation of the vehicle's engine; that is, check the air intake and exhaust. For example, a tank weighing 106,000 pounds is mired at fender depth. It cannot be recovered in the opposite direction of its original travel, but it can apply power to its tracks. Estimated resistance (twice the weight of the vehicle) is 212,000 pounds minus 40 percent. The load resistance equals 127,200 pounds.

Vehicle weight	106,000 lb
Mired factor (fender depth)	x 2
Resistance	212,000 lb
Reduction factor (40% for power to track)	x .40
Estimated reduction	84,800 lb
	212,000 lb
	- 84,800 lb
Estimated load resistance	127,200 lb

2-33. In another example, a tank weighing 106,000 pounds is mired at fender depth. If the tank can be recovered in the opposite direction of its original travel, and power can be applied to its tracks, the estimated resistance (twice the weight of the vehicle) is 212,000 pounds, less 50 percent (10 percent for opposite direction plus 40 percent for applying power to its tracks). The estimated load resistance equals 106,000 pounds.

Vehicle weight	106,000 lb
Mired factor	x 2
Resistance	212,000 lb
*Reduction factor	x .50
Estimated reduction	106,000 lb load
	212,000 lb
	-106,000 lb

Estimated load resistance	106,000 lb
*40 percent for applying power to track plus 10 percent recovering opposite direction to travel (50 percent)	

Note. Reduction factors do not apply to wheeled vehicles due to lack of traction. However, power applied to wheels may reduce resistance. Reduction factors are only a guide and apply more to wheel depth than to either fender or turret depth mire situations.

SOURCE OF EFFORT

2-34. Like vehicles are the quickest, most available sources of recovery effort. On dry, level hardstand in first gear or reverse, the average vehicle exerts a force equal to its own weight. Terrain conditions affect the towing capability of a vehicle. These conditions may require two or more vehicles to exert the same force that one vehicle normally could under ideal conditions. A winch is used when the situation does not permit recovery by a like vehicle. (Most often, the approach to the disabled vehicle does not provide good traction.) A winch is a more positive source of effort since its towing capability does not depend on terrain conditions.

WINCH VARIABLE CAPACITIES

2-35. A winch exerts its greatest force when it pulls by the first layer or the layer next to the bare winch drum. As each successive layer of cable is wound onto the winch drum, the diameter increases and winch capacity decreases.

2-36. An exception is the constant pull winch found on the M88A2—where the force of pull remains constant regardless of the cable layer. See table 2-1 for a listing of estimated winch variable capacities. Refer to the equipment's operator's manual for specified capabilities.

Table 2-1. Estimated winch variable capacity

Winch Type	*Cable Layer*	*Cable on Drum (Feet)*	*Capacity (Tons)*
5 ton	1	0 – 39	5.000
	2	40 – 85	4.225
	3	86 – 138	3.670
	4	139 – 199	3.230
	5	200 – 266	2.890
10 ton	1	0 – 41	10.000
	2	42 – 91	8.450
	3	92 – 148	7.250
	4	149 – 213	6.400
	5	214 – 287	5.700
22.5 ton	1	0 – 42	22.500
	2	43 – 93	18.850
	3	94 – 153	16.250

	4	154 – 220	14.250
	5	221 – 296	12.650
	6	297 – 380	11.400
30 ton	1	0 – 55	30.000
	2	56 – 128	26.000
	3	129 – 208	23.000
	4	209 – 300	20.000
45 ton	1	0 – 41	45.000
	2	42 – 91	38.000
	3	92 – 149	32.500
	4	150 – 200	28.500

Note. The 70-ton recovery vehicle has a constant capacity of 70 tons anywhere on the cable.

OVERCOMING RESISTANCE

2-37. Applying effort to overcome resistance has always been a challenge. Modern machinery makes this evident. Energy released by burning small amounts of fuel in a modern engine provides the effort to move vehicles weighing thousands of pounds. The vehicle engine, with various mechanical devices, can move the vehicle from a standstill through a wide range of speeds.

Mechanical Advantage

2-38. Mechanical advantage (MA) is a small amount of force applied over a long distance to move a heavy load a short distance. MA is needed whenever the load resistance is greater than the capacity of the available effort (AE).

2-39. To determine the amount of MA necessary in a recovery operation, divide the load resistance (LR) by the AE and round any fraction to the next whole number. Rounding is required because only whole numbers can be rigged.

$$\frac{\text{LR} = 106{,}000 \text{ lb (load)}}{\text{AE} = 90{,}000 \text{ lb (winch)}} = 1.1777$$

Round the fraction off to the next whole number = 2

Required MA = 2:1

Leverage Principle

2-40. Using levers is the most basic means to overcome resistance. A wrench handle and the gears of a truck overcome resistance by applying the principles of leverage. The simplest form of a lever is a rigid bar free to turn on a fixed pivot called a fulcrum. When effort is exerted on one end of the bar, the bar rotates around the fulcrum. MA is increased by extending the distance between the point where effort is applied and the fulcrum.

Lever Classification

2-41. The location of the fulcrum with relation to effort and resistance determines if it is a first-class or second-class lever.

First-Class Lever

2-42. The fulcrum is located between the effort and the resistance (figure 2-7). A crowbar is a good example of a first-class lever.

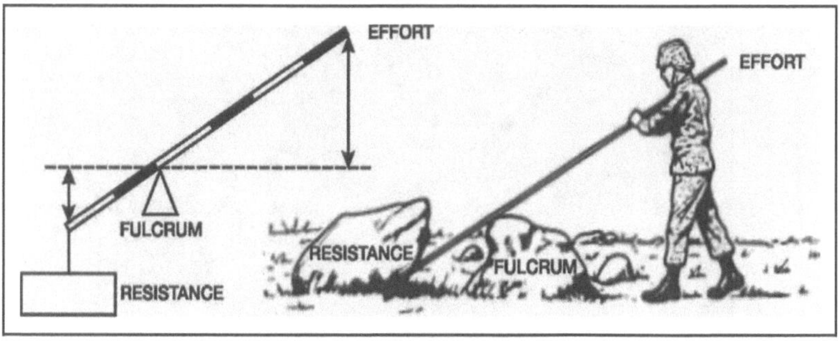

Figure 2-7. First-class lever

Second-Class Lever

2-43. The point of resistance is between the fulcrum and the effort (figure 2-8). A wheelbarrow is a good example of a second-class lever.

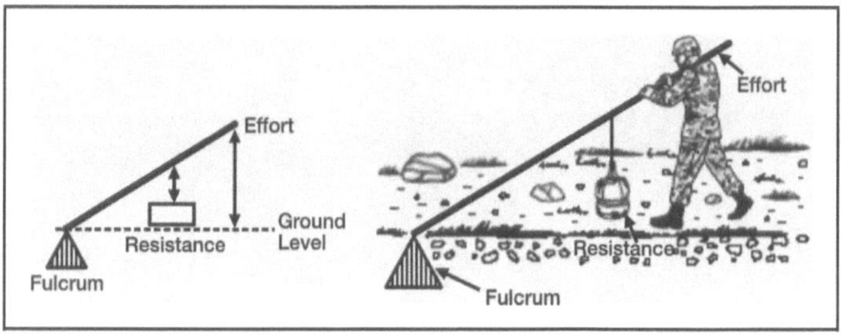

Figure 2-8. Second-class lever

Tackle Systems

2-44. Tackle is a combination of cables and blocks used to gain an MA or to change the direction of pull and are classified as either simple or compound.

Simple Tackle System

2-45. Simple tackle is one cable with one or more blocks.

Compound Tackle System

2-46. Compound tackle is a series of two or more simple tackles (figure 2-9). The output of one simple tackle is used as the effort for the other. Because a winch has only one cable, simple tackle will be used during most recovery operations.

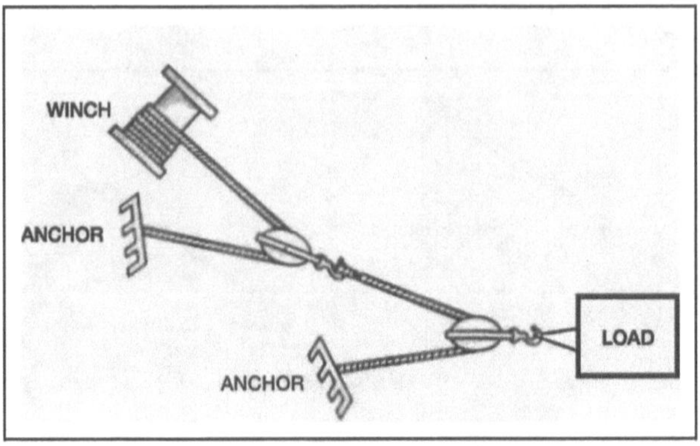

Figure 2-9. Compound tackle system

Chapter 3

Rigging

Rigging is applying wire rope in various tackle combinations to raise or move loads. Rigging involves installing the necessary equipment to use the AE, and it may or may not produce MA.

RIGGING FUNDAMENTALS

FALL LINE

3-1. The fall line is the winch line that runs from the source of effort to the first block in the tackle. There is only one fall line in a simple tackle system (figure 3-1). The amount of force that must be exerted on the fall line relative to the AE must be considered in every problem. The fall line force must be less than the capacity of the effort to accomplish the recovery. See the examples in paragraph 3-9, steps 1 through 4.

To determine the fall line force, divide the total resistance by the MA of the tackle.

$$\frac{\text{total resistance}}{\text{MA of the tackle}} = \text{fall line force}$$

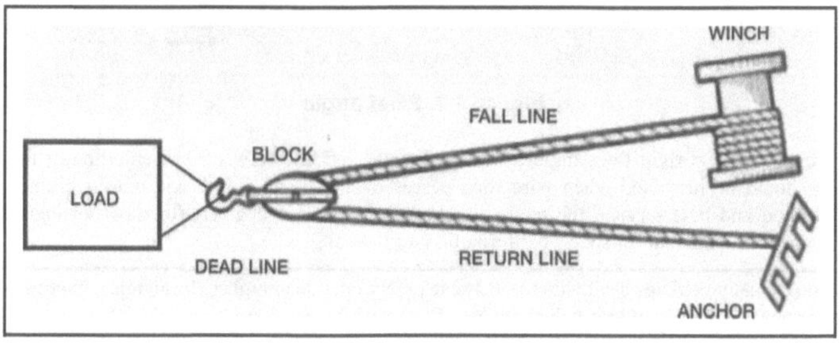

Figure 3-1. Terminology of simple tackle

RETURN LINE

3-2. A return line is a winch line rigged between the block or the winch line from the sheave of a block to the point where the end of the line is attached. This force is always the same as the fall line force (figure 3-1).

DEAD LINE

3-3. A dead line is a line used to attach blocks or other equipment to the load or to an anchor. To determine the dead line force, multiply the fall line force by the highest number of winch lines supported by the dead line (figure 3-1).

FLEET ANGLE

3-4. Achieving even winding of the winch cable on the drum is important for wire rope life and winch operations. This is best accomplished by working with the proper fleet angle.

3-5. Figure 3-2 displays the wire rope running from a fixed sheave, over floating sheaves, and then onto the surface of a smooth drum. The fleet angle is defined as the included angle between two lines.

- One line is drawn through the middle of the fixed sheave and the drum and perpendicular to the axis of the drum.
- A second line is drawn from the flange of the drum to the base of the groove in the fixed sheave.

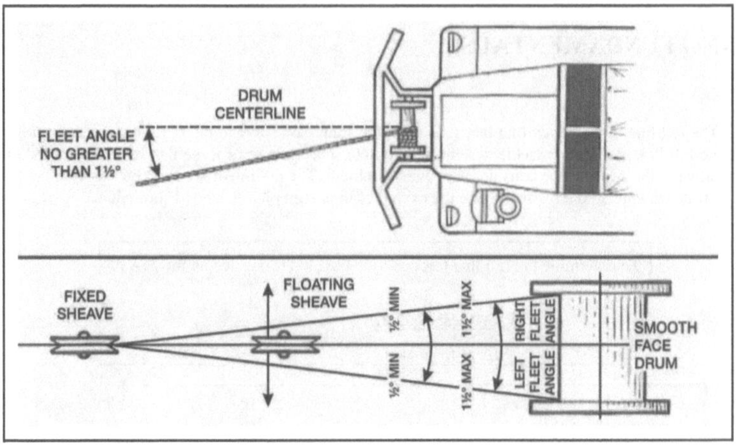

Figure 3-2. Fleet angle

3-6. There are left and right fleet angles, measured to the left and right of the centerline of the sheave. The fleet angle should be restricted when wire rope passes over a fixed sheave and onto a drum. For the most efficient method and best service, the angle should not exceed 1½ degrees, for most vehicles. Refer to the equipment operator's manual for specific information on fleet angles.

Note. Although many vehicles have winches that can safely operate at higher fleet angles, maximum stability and performance is achieved at lesser fleet angles.

MECHANICAL ADVANTAGE OF TACKLE

3-7. MA is needed whenever the load resistance is greater than the AE. The amount of MA needed is estimated by dividing the load resistance by the AE. The MA of any simple tackle system is equal to the number of winch lines supporting the load or the number of winch lines that become shorter as power is applied to the winch (figure 3-3). The lines can be attached directly or indirectly through a block.

3-8. Placement of the block is critical to gaining MA. The block must be attached to the movable load and effort applied in the opposite direction to divide the effort equally over the two lines. The arc over the lines indicates lines that support the load (figure 3-3).

Note. The 1-to-1 ratio shown has an arc over only one line indicating the block is simply changing the direction of effort. No mechanical advantage is gained in this configuration.

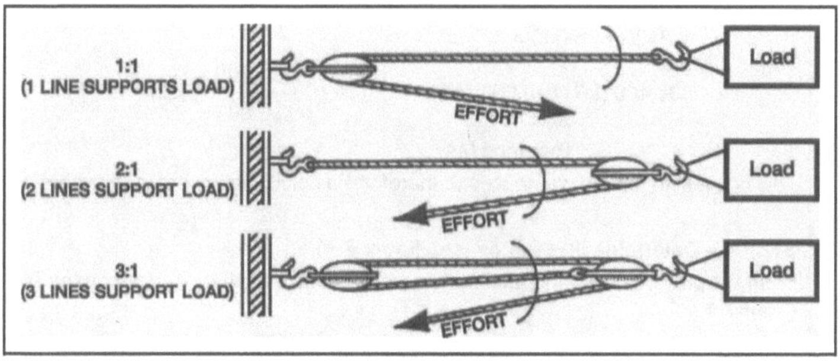

Figure 3-3. Winch line(s) MA

DETERMINING LINE FORCES

3-9. The following example shows how to compute various line forces. A disabled vehicle had a load resistance of 14 tons (28,000 pounds). The AE is a winch with a maximum capacity of 5 tons (10,000 pounds). What MA must be rigged to recover this vehicle? What are the line forces?

STEP 1 – Determine initial estimate.

Load resistance / AE $\frac{LR}{AE}$ = Required MA

28,000 (LR) / 10,000 (AE) $\frac{28,000}{10,000} = 2.8$

Required MA = 3:1

STEP 2 – Add tackle resistance and verify solution.

An MA of 3 requires 2 sheaves. To determine the tackle resistance, multiply 10 percent (.10) of the load resistance by the number of sheaves.

(.10 x 28,000) x 2 = 5600

Then add the tackle resistance to the load resistance for total load resistance.

5600 + 28000 = 33,600

33,600 (LR) /10,000 (AE) $\frac{33,600}{10,000}$ = 3.36

Required MA = 4:1

This is not equal to the answer in step 1; therefore, the answer must be re-verified.

STEP 3 – Re-verify solution.

An MA of 4 requires 3 sheaves. Therefore, 3 x .10 (10% per sheave) or .30 (30%) must be added to the load resistance.

$(.10 \times 28{,}000) \times 3 = 8400$

$8400 + 28{,}000 = 36{,}400$

36,400 (LR) / 10,000 (AE) $\frac{36{,}400}{10{,}000} = 3.64$

Required MA = 4:1

This is equal to the answer in step 2; therefore, a solution has been achieved.

STEP 4 – Determine line forces (see figure 3-4).

From the previous step, the total resistance is equal to 36,400 pounds and the MA needed is 4:1.

Total resistance (LR) / MA = Fall line force

36,400 / 4 = fall line force $\frac{36{,}400}{4} = 9{,}100$ lb

Note. Double check: The fall line force is less than the winch capacity.

Return line force	= Fall line force
Fall line force	= 9,100 lb
Return lines 1, 2, and 3	= 9,100 lb each

The dead line force equals the number of support winch lines times the fall line force. (See figure 3-4.)

Lines 1, 2, 3 (and fall line)	= 9,100 lb	
Dead line I	= 4 x 9,100 lb	= 36,400 lb
Dead line II	= 2 x 9,100 lb	= 18,200 lb
Dead line III	= 1 x 9,100 lb	= 9,100 lb

Note. Ensure Y-slings used for dead lines in this rigging operation are rated to withstand the force applied at the point of attachment.

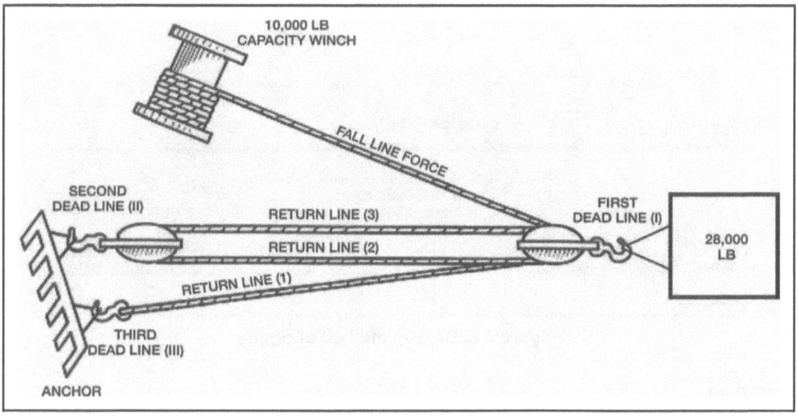

Figure 3-4. 4-to-1 MA

Note. If field expedient slings are used as dead lines, refer to FM 5-125 to determine sling leg forces. Field expedient slings are considered slings that are constructed using materiel not part of the recovery vehicle's BII.

METHODS OF RIGGING

3-10. The rigging techniques used depend on terrain, the type of vehicle, and the distance between the recovery vehicle and the disabled vehicle. Manpower, backup, and lead methods for rigging techniques are discussed below.

Manpower Method

3-11. The manpower method is used when the winch cable and other rigging equipment are lightweight and can be carried easily by the crewmembers to where they are needed. This method depends completely on Soldier manpower.

Backup Method

3-12. The backup method is used when the recovery vehicle can be safely positioned within 20 to 25 feet of the disabled vehicle. Figure 3-5 shows the recovery vehicle in position to perform the winching operation.

- Pull out enough main winch cable to attach to the disabled vehicle.
- Place the main winch snatch block in the loop of the cable and attach the block to the disabled vehicle.
- Back up the recovery vehicle, allowing the main winch cable to be spooled from the winch drum until sufficient cable is removed to obtain maximum winch capacity.

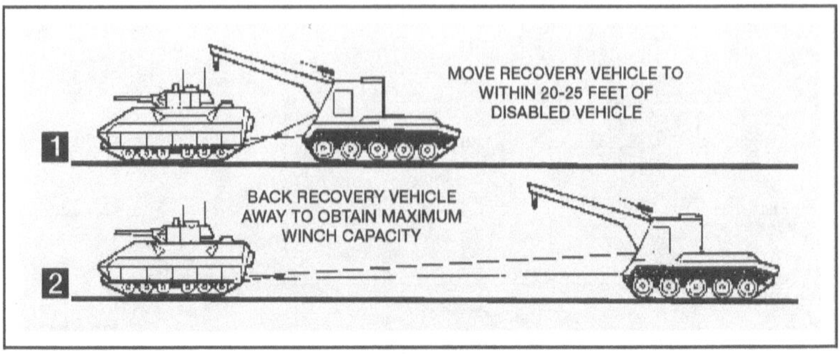

Figure 3-5. Backup method of rigging

Lead Method

3-13. The lead method (figure 3-6) is used when terrain conditions do not permit close access to the disabled vehicle. Use the boom winch or auxiliary winch cable to pay out the main winch rigging to the disabled vehicle. Since the boom or auxiliary winch cable weighs less than the main winch cable, it can easily be carried to the disabled vehicle.

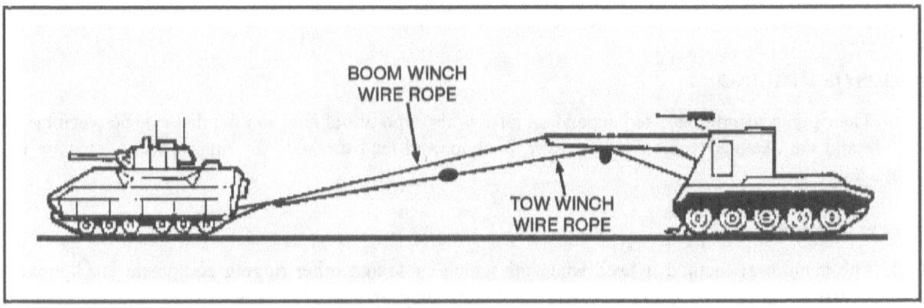

Figure 3-6. Lead method of rigging

3-14. To rig for the lead method, assemble the main winch tackle just in front of the recovery vehicle as in preparation for the backup method. Then do the following:

- Attach the boom and crane or auxiliary winch cable to the main winch snatch block.
- Manually pull out the loop formed by the boom and the crane or auxiliary winch cable, and place it into a snatch block attached to the disabled vehicle.
- Lastly, attach the cable to the rigging. (By paying in the boom and the crane or auxiliary winch cable, the main winch tackle will be pulled to the disabled vehicle.)

METHODS OF ATTACHING TACKLE

Wheeled Vehicles

3-15. In recovery operations, rig the tackle to prevent damage to the vehicle or equipment. For instance, on disabled wheeled vehicles, attach the rigging (tackle) to the bumper lifting shackles on both sides of the tow pintle. If the pulling force is attached only to one frame member, the truck frame could be pulled out of alignment.

CAUTION

Even though the bumper lifting shackles can withstand force from a horizontal or vertical pull, a sling attachment must be used to apply the effort equally to both shackles. For vehicles not equipped with bumper lifting shackles, effort should be applied to the main structural members, not to the bumper or bumper brackets

3-16. On wheeled vehicles, whether the pull is made from the front or rear, apply the effort to both of the bumper lifting shackles (figure 3-7). The force exerted on each leg of the sling is slightly greater than half of the resistance.

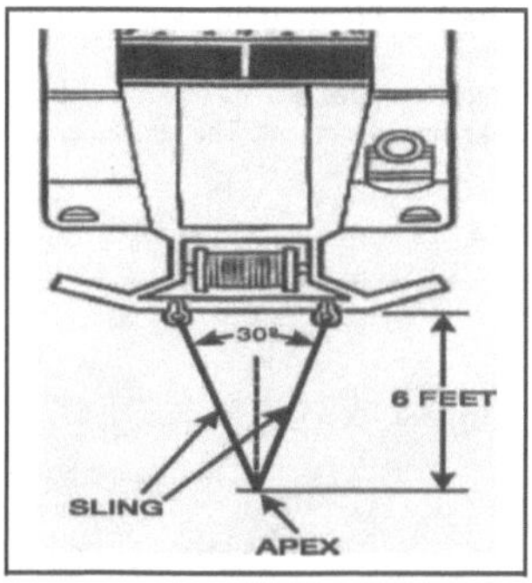

Figure 3-7. Sling arrangement

Tracked Vehicles

3-17. On tracked vehicles, always attach rigging to the tow hook or lugs. The lifting eyes are not designed to withstand the pulling force required for recovery. Use an attachment that will distribute the applied force to each side of the vehicle.

3-18. When a disabled tracked vehicle does not require towing or MA, use the main winch snatch block with one tow cable to form a floating block hookup. This hookup is easy to install and distributes the effort evenly to both tow hooks. To rig a floating block—

- Attach the ends of the tow cable to the two tow hooks.
- Place the snatch block in the loop formed by the tow cable.
- Attach the winch cable to the snatch block.
- Ensure that cables and attachments can withstand forces as shown in figure 3-8.

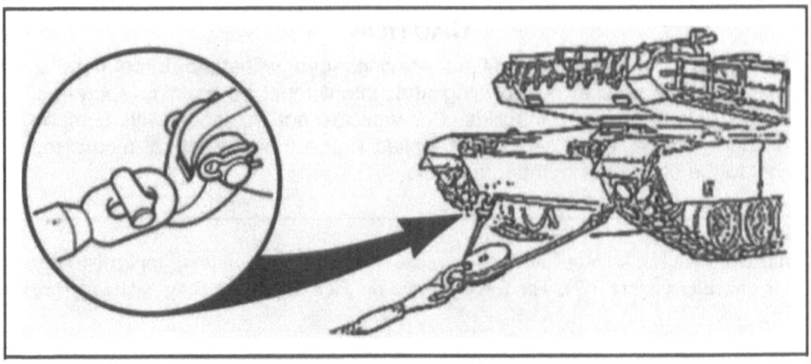

Figure 3-8. Floating block attachment

3-19. When a disabled tracked vehicle requires a 2-to-1 MA rigging and towing over rough terrain, after winching, use two tow cables to make the attachment. The attachment, shown in figure 3-9, is the quickest to rig.

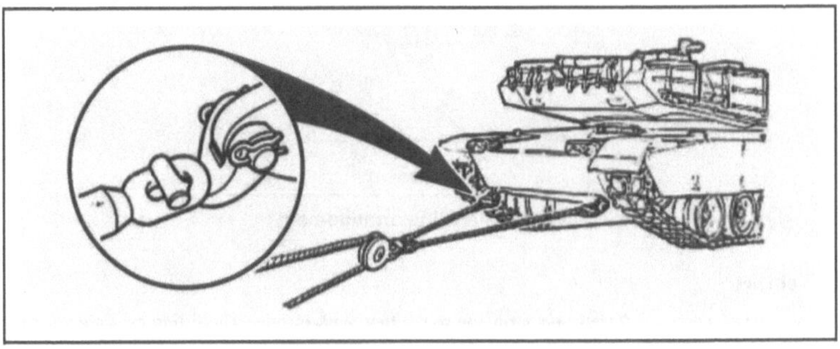

Figure 3-9. Tow cable attachment

3-20. When towing a tracked vehicle over relatively level terrain or on highways, use the tow bar method of attachment (figure 3-10). Attach the tow bar to the tow lugs of the disabled vehicle and attach the winch rigging to the lunette of the tow bar. After winching, disassemble the rigging and place the tow bar lunette in the recovery vehicle's tow pintle.

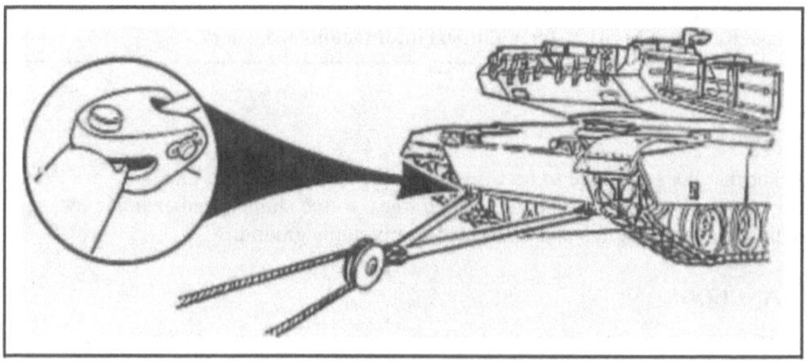

Figure 3-10. Tow bar method

3-21. If a 3-to-1 MA is required, attach the running block to one of the tow lugs of the disabled tracked vehicle, attach the change-of-direction block to the tow lug on the recovery vehicle, and attach the end of the winch cable to the other tow lug on the disabled tracked vehicle. This method can be used in the absence of tow cables but a holdback vehicle is required. This is a time-consuming method in the absence of a holdback vehicle because only one vehicle will move at a time. When the recovery vehicle is moved, the disabled tracked vehicle will have to be chocked to prevent it from moving. Figure 3-11 shows the 3-to-1 MA towing method.

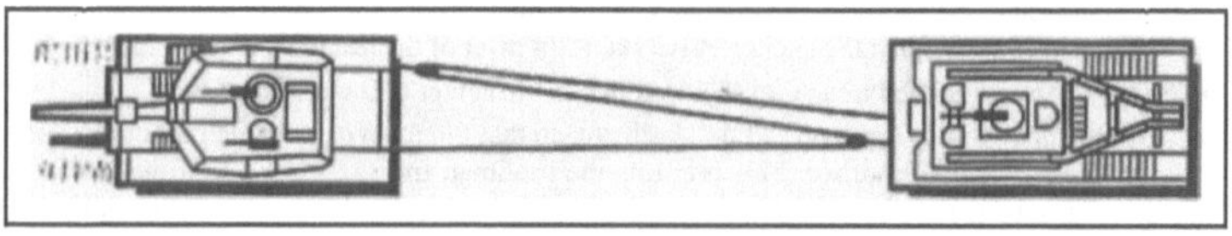

Figure 3-11. 3-to-1 MA towing method

ANCHORS

USE OF ANCHORS

3-22. Anchors are used to create a solid point of attachment for connecting rigging during recovery operations. Multiple anchors provide additional points of attachment for rigging mechanical advantage, add resistance to the effort, and prevent vehicle slide when recovering on a slope. Frequently, wheeled and tracked vehicles must have some anchoring means when winching heavy loads with tackle (figure 3-12 shows the anchor symbol). An anchor can assist in holding a recovery vehicle, improvising a change of direction pull, or supporting part of the load during a winching operation. Most existing recovery vehicles have ground chocks or spades to provide resistance while performing recovery; however, additional anchoring may be required if the recovery vehicle continues to slide when the winch is operated.

Figure 3-12. Anchor symbol

Note. Trees, tree stumps, large rocks, or other vehicles may be used as anchors in recovery operations. Refer to FM 5-125 for additional information on anchors.

NATURAL ANCHORS

3-23. An anchor that does not have to be constructed is a natural anchor. Examples are trees, tree stumps, and large rocks or other vehicles. Avoid dead or rotten trees or tree stumps, and examine rocks and trees carefully to make sure they are large enough and embedded firmly in the ground.

MECHANICAL ANCHORS

3-24. There are several types of mechanical anchors. The anchor type used/ constructed depends on holding ability requirements, the type of soil required, the availability of material, and the situation.

Log Deadman

3-25. A log deadman (figure 3-13) is one of the best types of anchors for heavy loads. The deadman consists of a log buried in the ground with the dead line connected to its center. When constructing a deadman—

- Place the deadman where the direction of pull is as horizontal as possible. Take advantage of sharp banks or crests to increase the holding power with less digging.
- Dig a trench large enough for the deadman and as deep as necessary for good bearing. When digging, slant the trench in the direction of the pull at an angle of approximately 15 degrees from the vertical. To strengthen the anchor, drive stakes in front of the deadman at each end.
- Dig a narrow inclined trench for the dead line at the center of the deadman.
- Tie the dead line to the center of the deadman, so that the main or standing part of the line leads from the bottom of the deadman. This prevents the deadman from rotating out of the trench. If the dead line has a tendency to cut into the ground, place a small log under the line at the outlet of the trench. The strength of the deadman depends on the strength of the log and the holding power of the earth.

Note. The procedures that follow in this chapter require an understanding of specific hitches and knots. Refer to FM 5-125, chapter 2, for further explanations and illustrations of hitches and knots.

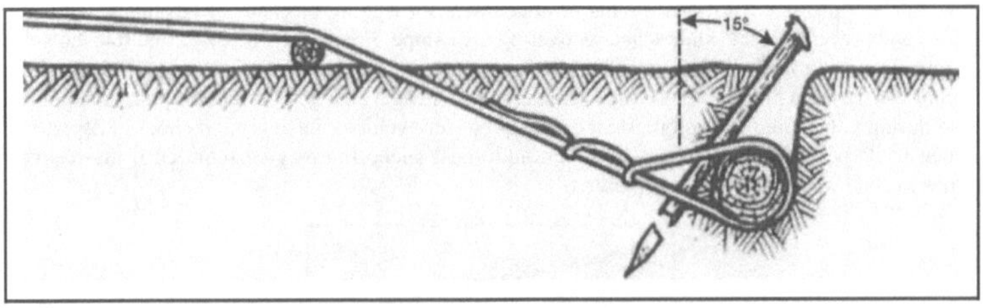

Figure 3-13. Log deadman

Picket Holdfasts

3-26. A picket holdfast (figure 3-14) is constructed by using two or more sound wooden pickets at least 3 inches in diameter and 5 feet long.

- Drive the pickets approximately 3 feet into the ground, 3 to 6 feet apart, and in line with the dead line.
- Tie a pair of pickets together.
  - With fiber rope, tie one end of the rope to the top of the front picket with a clove hitch.
  - Then make four to six wraps of the rope from the top of the first picket to the bottom of the second picket, ending with a clove hitch to the bottom of the second picket.
- Insert a stake between the rope wraps midway between the pickets.
  - Tighten the rope by twisting it with the stake.
  - Drive the stake into the ground.
- Repeat this procedure for each successive pair of pickets.

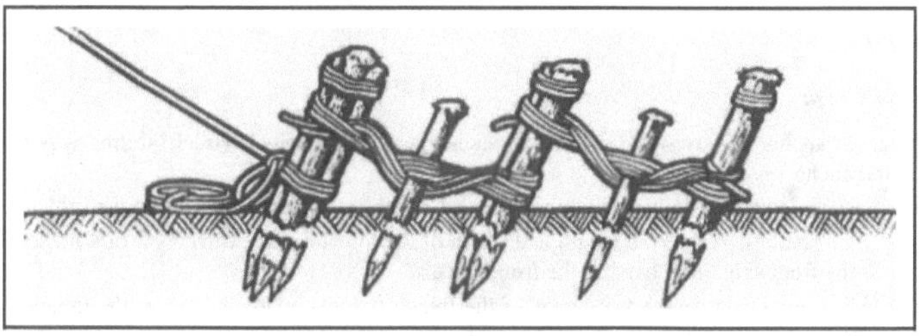

Figure 3-14. Combination picket holdfast

Note. The strength of the holdfast depends on the strength of the first or front picket. To reinforce it, drive two or more pickets into the ground close to the front picket. Tie all pickets in the front picket together using a clove hitch; wrap the front picket to the second picket, insert the stake, and drive the stake into the ground.

Sand Parachute

3-27. A sand parachute (figure 3-15) can be used in sandy areas when trees or other attachable devices are not available. Construct a sand parachute as follows:

- Dig a hole, line the hole with a tarpaulin, and fill the hole with sand removed from the hole.
- Lash the four corners of the tarpaulin together with rope and attach rigging to the rope.

Note. Multiple sand parachutes can be used in tandem by attaching them to a chain or Y-sling. This will provide additional holding strength, but the sand parachute has a limited holding ability. A sand parachute should not be used when a major effort is required.

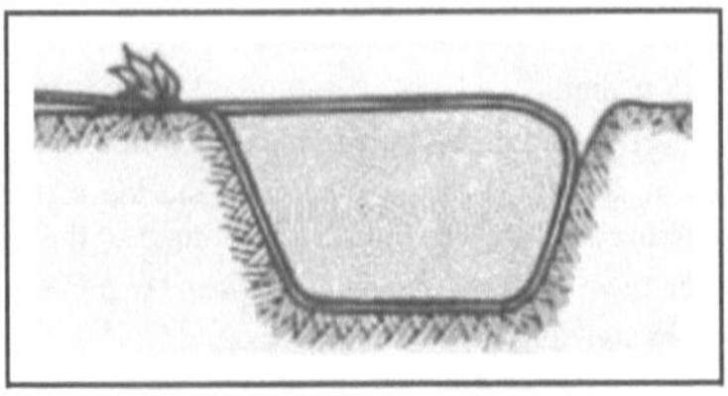

Figure 3-15. Sand parachute

Scotch Anchor

Using One Tow Chain

3-28. A scotch anchor (figure 3-16) is used to anchor a wheeled vehicle (truck) during winching operations when natural anchors are not available.

- Select a log at least 6 inches in diameter and 2 feet longer than the width of the vehicle.
- Dig a shallow trench (the length and width of the log and approximately 3 or 4 inches deep) parallel to the front axle, just ahead of the front wheels.
- Lay a tow chain across the center of the trench (width), place the log in the trench, and move the vehicle forward until both front wheels are against the log.
- Attach both chain ends to the bumper lifting shackles and remove all slack from the chain.

Note. As pressure is applied to the winch, the front wheels are pulled onto the log, making the chain taut and anchoring the vehicle.

Figure 3-16. Scotch anchor

Using Two Tow Chains

3-29. If more than one tow chain is available, a similar method may be used.

- Lay two tow chains across the center of the trench (width) next to the inside of each front wheel.
- Place the log in the trench and move the vehicle forward until both front wheels are against the log.
- Wrap the chains through the bumper lifting shackles, remove the slack from the chains, and fasten them together using chain hooks.

Vehicle

3-30. A vehicle can be used as an anchor to assist in the recovery of a mired vehicle equipped with a winch. The winch cable from the mired vehicle is payed out to the anchoring vehicle and the mired vehicle winches itself out. The anchoring vehicle should not attempt to pull; it is only an anchor.

Chapter 4
Recovery Operations

A successful recovery operation is done quickly and safely. Care must be exercised when erecting and using equipment to prevent damage to vehicles and equipment and to prevent injury to personnel.

CAUTION

Think Safety

Recovery is a big job. Before any recovery operation, calculate the resistance, inspect tackle, and keep rigging references handy. A haphazard approach to recovery can lead to dismemberment, death, and/or damaged equipment.

SAFETY PRECAUTIONS

ACCELERATION IMPACT

4-1.Failure occurs when a weight is allowed to fall for a distance and is suddenly stopped. A similar strong force occurs when power is engaged suddenly to recovery vehicles when connected to a towed or mired vehicle. Do not apply loads suddenly (shock load). This puts excessive strain on the equipment, and it may fail.

CAUTION

A winch line makes a deadly slingshot. If the dead line of a snatch block breaks, a 200-pound snatch block can travel as far as 300 yards in the air. All personnel observing should stand at least one cable length away from (or the length designated by the recovery vehicle) and opposite of the angle of pull (figure 4-1) when the cable is under stress. This will allow greater reaction time for personnel to move out of the path of flying objects if a cable or other attaching hardware breaks.

BACKLASH

4-2.Make every effort to stand clear of wire rope that is under tension. The minimum safe distance is twice the length of the payed-out cable. When wire rope is drawn taut and then released suddenly by a break, its recoil (or backlash) may cut a person into two pieces. A winch line under load stretches like a rubber band and stores up tremendous potential kinetic energy. In fact, a steel winch cable weighing 50 to 500 pounds has more spring to it than rubber.

Crossed Cables

4-3.Make sure the rigging lines are not crossing each other before the winching operation is continued. Crossed rigging lines can rub against each other causing damage to the cable or an increased amount of tackle resistance. Crossed cables are only recommended for towing a disabled vehicle when a tow bar is not available.

Ground Guides

4-4.To safely control a recovery operation, use two ground guides—one ground guide in the front and one in the rear. Only one ground guide gives the signals to the operator. The ground guides should stand apart from other personnel at the recovery site and be in a position where the vehicle operators can easily observe the signals. The vehicle operators must know the meaning of the signals and act only on those signals (see appendix C).

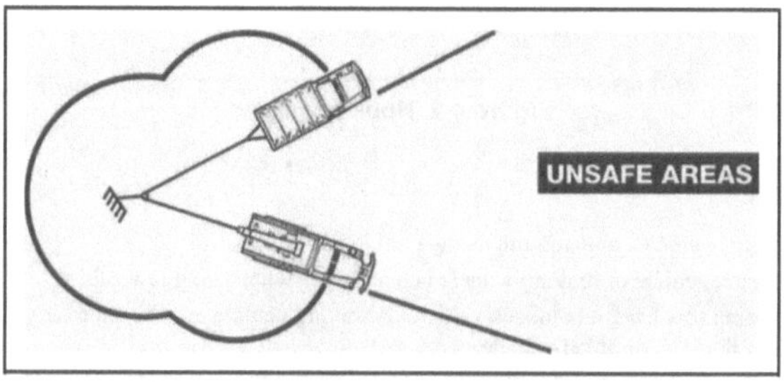

Figure 4-1. Unsafe areas during an angle pull

Hook Positions

4-5.For rigging, position the hook with the open part (throat) upward (figure 4-2). If the hook should straighten out from overload, the rigging would be forced downward. If the hook were positioned with the open part (throat) down, the rigging would travel upward unrestricted and possibly cause injury to personnel or damage to vehicles.

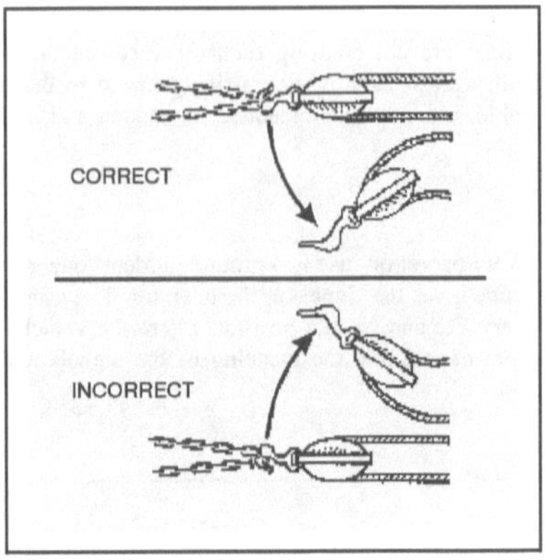

Figure 4-2. Hook positions

Holdback Vehicles

4-6. Towing tracked vehicles may require using a holdback vehicle.

- A holdback vehicle or braking vehicle is mandatory when using tow cables.
- When using tow bars, a holdback vehicle or braking vehicle may be necessary if the recovery vehicle is lighter than the disabled vehicle.

CAUTION

Inspect rigging thoroughly at every connection to ensure that safety pins are installed correctly and that proper shackles, pins, and hooks are used. Ensure that tow cables are not crossed and are reeved correctly in the snatch block.

Inspecting Rigging

4-7. Inspect equipment thoroughly before the recovery operation starts. Direct the recovery vehicle operator to apply power to the winch to remove the slack from the rigging, and then stop the operation so the rigging can be inspected without endangering personnel. When inspecting the rigging, never place the hands or body between cables under tension.

Operator/Driver Safety

4-8. Operators and other personnel, in both the recovery and disabled tracked vehicles, must keep their hatches closed during winching and AKERR operations. Operators should use their periscopes to view hand and arm signals.

Positioning Gun Tubes

4-9. During tank or tracked howitzer recovery, position the main gun tube so it will not be damaged. If the gun tube of a disabled tank or tracked howitzer is involved in a collision (this might occur on a nosed or overturned tank), maintenance support personnel should always check the gun tube before firing.

Rigging Between Vehicles

4-10. While rigging is being erected between vehicles, turn off the engines and apply the brakes. This prevents possible injury to recovery personnel and/or damage to the vehicles. When riggings are erected using a recovery vehicle that must have its engine running to operate the equipment, position the spade or chocks (wheeled vehicle) and apply the brakes to prevent movement (figure 4-3). The driver remains in position.

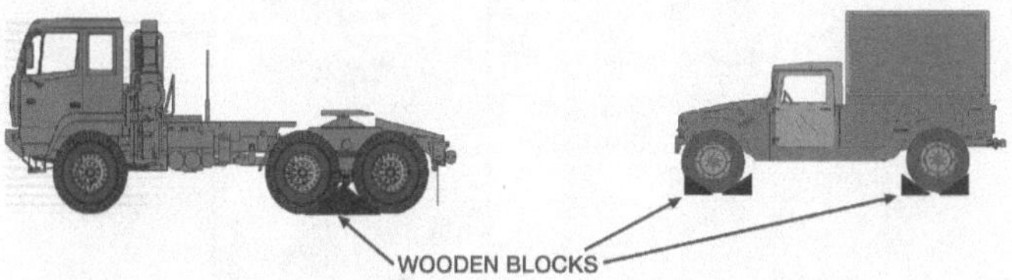

HOOKING UP TO A VEHICLE:
Before hooking up to the tow bar or before disconnecting the drive or parking brake, chock the vehicle wheels with blocks so that it cannot move. (Place a block of wood or other suitable object between the rear tires or in the front of and in back of one tire.) Make sure the chock extends the full width of the tire.

UNHOOKING FROM A VEHICLE:
Before unhooking the tow bar or before hooking up the drive or parking brake, chock the vehicle wheels with blocks so that it cannot move. (Place a block of wood or other suitable object between the rear tires or in the front of and in back of one tire.) Make sure the chock extends the full width of the tire.

CAUTION
Failure to remove the blocks could result in damage to the vehicle.

Figure 4-3. Chocking/blocking wheeled vehicles

Safety Keys and Shackle Pins

4-11. Safety keys/shackle pins should be in place on all tow hooks, shackles, or other items of equipment (figure 4-4). Even though the safety key/shackle pin supports no great load, its absence can allow a pin to move which places excessive force on only a part of a connection. Some shackles use a threaded-type pin. If the pin is not completely inserted into the shackle threads, the shackle or pin can be bent or broken when force is applied.

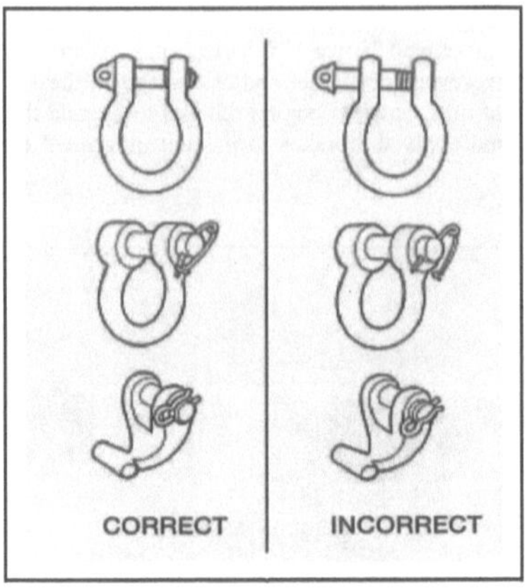

Figure 4-4. Safety keys and shackle pins

4-12. When using shackle pins with safety keys, such as the type used in tow bars, all shackle pins in a vertical plane should have their heads pointing upward. Should the safety key break or fall out, the shackle pins will remain in position if the load shifts.

SPEED

4-13. Maintain the correct speed when towing vehicles. Consider the terrain, weather, and road conditions when determining speed. Never exceed the tow speeds listed in the operator's manuals for the towed or towing vehicle.

RECOVERY PROCEDURES

4-14. In any recovery operation, use the following eight-step method.

Reconnoiter area
Estimate situation
Calculate ratio
Obtain resistance
Verify solution
Erect rigging
Recheck rigging
You are ready

STEP 1. Reconnoiter area.

- Check the terrain for the best approach to the load; then determine the method of rigging and the availability of natural anchors. A recovery crew must know the problem before making decisions.
- Conduct a complete ground survey of the area; then select the best route of approach to the disabled vehicle to prevent possible disablement of the recovery vehicle. When selecting the evacuation route, ensure that the military route classification number will support the combination vehicle classification (recovery plus towed vehicles). Refer to FM 5-170 for further information.

STEP 2. Estimate situation.

- Estimate the resistance created by the load and determine the capacity of the AE. For most recovery operations involving winching, the AE would be the maximum capacity of the winch. In some recovery operations, the maximum distance between the winch and the disabled vehicle could be restricted, making the AE as little as half of the winch capacity.

STEP 3. Calculate ratio.

- Compute an estimated MA for the rigging by dividing the resistance of the load (step 2) by the AE (the capacity of the winch).

STEP 4. Obtain resistance.

- Compute the tackle resistance and total load resistance. Multiply the percent of the load resistance (as determined in step 2) by the number of sheaves in the rigging. The determined resistance of the tackle added to the load resistance equals the total resistance.
- Total effort available is winch capacity multiplied by the MA (as computed in step 3). If effort available is more than total resistance, proceed to step 5. If it is less, return to step 3 and add MA.

STEP 5. Verify solution.

- Compute line forces to compare with the winch and dead line capacities. Divide the total load resistance (step 4) by the MA (step 3). The result is the fall line force. The fall line force must be less than the capacity of AE. Therefore, this step of the recovery procedure is the key step to solving the problem.

Note. When verifying the solution, if the computed fall line force is greater than the effort, the MA must be increased. Note that no physical work has occurred up to this point. As a result, no time is lost moving equipment or having to re-erect rigging equipment.

- Compute the dead line force, determine the required strength of equipment capacity, and choose the correct equipment to use as dead lines.

STEP 6. Erect rigging.

- Orient the crew, instruct them on how to assemble the tackle, and then move to a safe location.
- Advise the crewmembers of the plan, direct them to erect the tackle, and assign specific tasks for desired MA. (Crewmembers that have finished their tasks should assist those who are having difficulty. The crewmembers can save time by having a thorough knowledge of the tackle to be erected and by helping each other.)
- Observe all safety precautions.

STEP 7. Recheck rigging.

- Make sure that the tackle is rigged for proper and safe operation.
- Direct the operator to remove most of the slack from the lines and to inspect for correct assembly. If any corrections must be made, direct the crewmembers to make them.
- Explain the details of the operations to the operators of the recovery vehicle and the other vehicles involved. Direct operators to watch for signals and be prepared to act on them. Then move to a safe location where signals can be observed by all vehicle operators.

STEP 8. You are ready.

- Signal the operators to apply winch power and recover the load. Be alert; make sure that nothing obstructs the operations of the equipment and that all personnel on the ground remain at a safe location.

Recovery Methods Using Wheeled Recovery Vehicles

Types and Usage

4-15. Trained recovery personnel must perform recovery operations (see appendix D). Trained recovery personnel use special purpose vehicles for recovery when methods used by the operator, crew, or platoon do not fit the situation or when their efforts have had no success. The methods of recovery performed with special purpose vehicles are winching, lifting, and towing.

Note. This section summarizes winch, lift, and tow procedures. For more in-depth information, refer to the equipment operator's manual, which relates to the operation of the equipment and its specific capabilities.

Winching

4-16. During the recovery of a mired truck using a wrecker, consider the following factors: the resistance of the load, the approach to the load, and the distance between the wrecker and mired vehicle. Mired trucks may have a resistance greater than the winch capacity of the wrecker. Also, the wrecker may not be able to align itself with the truck because of the terrain. If so, use a 2-to-1 MA and a change of direction block (figure 4-5).

Figure 4-5. Winching using a 2-to-1 MA and a change of direction block

Wheeled Towing

4-17. Recovered vehicles must often be towed to a repair shop. The towing method depends on the terrain and the mechanical condition of the disabled vehicle. A wrecker is capable of towing vehicles in the following ways.

Highway Tow

4-18. To use a highway tow—

- Attach the tow bar to the disabled vehicle's lifting eyes and the wrecker's tow pintle (figure 4-6). All wheels of the towed vehicle are on the ground. (A driver is not required in the towed vehicle.)
- Determine terrain features prior to travel. If the hill is steep enough to require using brakes to reduce speed, shift into the next lower gear at the crest of the hill, and use the engine compression as a brake.
- Take extreme care to prevent excessive engine speed while descending a hill. Determine the suitable gear and shift, if necessary, at the crest of the hill before speed has increased from downhill movement. Ordinarily the gear required to ascend a hill is proper to use to descend it. Refer to the vehicle operator's manual for additional information.

Figure 4-6. Highway towing

CAUTION

Safety chains must be used in addition to the tow bar. Properly used, safety chains will retain a towed vehicle should the tow bar fail or become disconnected. Cross the chains under and around the tow bar. Fasten it to the shackles of the towing vehicle around a structural member or the underside of the vehicle to be towed. Leave sufficient slack for turns, but not enough to encounter road surface.

Cross-Country Tow

4-19. Over rough terrain, a cross-country tow (figure 4-7) has better control of the towed vehicle. To rig for the cross-country tow—

- Attach a chain-lifting sling or the hoisting bar between the truck's front lifting shackles.
- Attach a tow chain from the wrecker tow pintle to the disabled truck's lifting shackles.
- Place the hoist block hook in the lifting sling approximately 12 to 18 inches off the ground.

Figure 4-7. Cross-country towing

4-20. Extend the boom as little as possible to remove the slack from the tow chain to keep the towed vehicle from ramming into the rear of the wrecker truck. Support the boom with the shipper braces to prevent impact loads on the crane mechanism.

Lift-Tow

4-21. If the front end of the vehicle is damaged, use the lift-tow procedure even when the disabled vehicle is being towed on the highway.

- Attach the axle clamps to the front axle of the disabled vehicle.
- Attach the tow bar to the axle clamps.
- Be careful not to damage the brake lines.

Note. Procedures are the same as cross-country towing except that a tow bar is used in place of the chain.

WARNING

Extreme caution should be exercised to avoid damage to the towed vehicle. If possible, use the lift-tow procedure before using cross-country tow or highway tow procedures. Refer to the equipment operator's manual for lift-tow restrictions, vehicle preparation, precautions that must be taken, and vehicle speed when performing lift-tow operations.

Recovery Methods Using Tracked Recovery Vehicles

Winching

4-22. One recovery vehicle is used for recovering most mired tracked vehicles. To prepare for winching, position the recovery vehicle in line (as much as possible) with the mired vehicle.

Variable Winch

4-23. When recovering a mired tracked vehicle with a recovery vehicle that has a variable type winch, the maximum winching capacity is gained when the cable is almost fully payed out or has less than three wraps remaining on the drum. Always refer to the operator's manual for guidance on achieving maximum winch effectiveness.

Constant Pull Winch

4-24. Vehicles with constant pull winches can be positioned as close as practicable to the mired vehicle. Always allow distance for the mired vehicle to get on solid ground. Testing has shown that the flat, smooth hull of the Abrams-series tank provides less resistance than expected. This should allow most of the mired tanks to be recovered using a single line pull. Recovery with two recovery vehicles (figure 4-8) is used only when the load resistance of a mired tracked vehicle is so great that the calculated fall line force is more than the winch capacity of one recovery vehicle with a 3-to-1 MA.

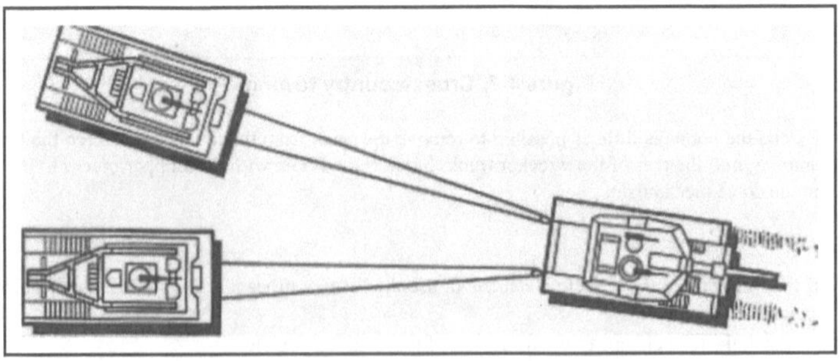

Figure 4-8. Winching with two recovery vehicles

4-25. To take full advantage of their winch capacities, the recovery vehicles are positioned side by side (figure 4-8). The same length of winch cable can then be used. Rig each recovery vehicle for a 2-to-1 MA (figure 4-9). Attach each rigging snatch block to a tow lug on the mired vehicle. To synchronize winch speeds, both recovery vehicle operators should use the hand throttle to set the engine speed at the desired revolutions per minute (RPMs) and compensate with the winch control lever to maintain tension on cables.

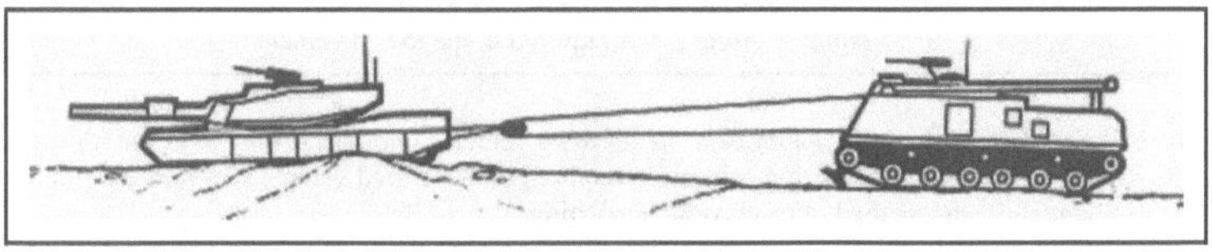

Figure 4-9. 2-to-1 MA

Tracked Towing

4-26. Although towing can be done with similar vehicles, it is often necessary for the recovery vehicle to tow a disabled vehicle to the UMCP where repairs can be made or evacuation effected. The towing method used will depend on the type of terrain and available assets.

Note. An observer will be used to assist the operator during towing operations. The observer will be located on the recovery vehicle and will have direct communication with the operator. The observer's responsibilities are to verify the tow connection throughout towing operations and alert the driver of unsafe conditions of the towed vehicle, including but not limited to, disconnect and/or jack knife. This may require that the recovery vehicle halt occasionally for the observer to exit the vehicle and examine the condition of the rigging. **At no time will the observer ride on the exterior of the recovery vehicle or the towed vehicle during towing operations.** The observer must be positioned to use observation windows and rearview mirrors.

Towing with the M88A1

4-27. When towing an Abrams family of vehicles (FOV) or M88A1 series recovery vehicle, always use another M88A1 or M1 series recovery vehicle as a holdback vehicle–even when using a tow bar.

- Never allow anyone to ride in or on an M1 while it is being towed.
- Never make sharp turns in first gear; make gradual wide turns.
- Never make sudden stops.
- Take care to protect Soldiers from exhaust heat and from skin contact with hot tow bars when hooking and unhooking tow devices.

Towing with the M88A2

4-28. There are operational restrictions when towing an Abrams FOV.

- Use an M88A2 recovery vehicle or, if dictated by the operational risk assessment, use another Abrams FOV as the braking or holdback vehicle.
- Use crossed tow cables as the method for attaching the braking or holdback vehicle.
- Use the tow bar provided with the M88A2 for the tow vehicle.
- When towing cross-country under extreme conditions, use crossed cables for the towing and braking or use a holdback vehicle as described in paragraph 4-31.

Highway Tow

4-29. For highway tows—

- Attach the recovery vehicle's tow bar to the tow lugs of the disabled vehicle.

- Place the lunette of the tow bar in the recovery vehicle's tow pintle. (This could be done using a small block and tackle—attaching one part to the tow bar and the other to a place on the recovery vehicle higher than the tow pintle. This will allow a single Soldier to raise the tow bar to the pintle without getting between the two vehicles.)
- Secure the pintle in the closed position.

Note. A driver is not required in the towed vehicle.

4-30. If the recovery vehicle is lighter than the disabled vehicle, a holdback vehicle of the same weight class as the disabled vehicle and cross tow cables are required so the towed vehicle will not overrun the recovery vehicle. Figure 4-10 shows the holdback vehicle positions.

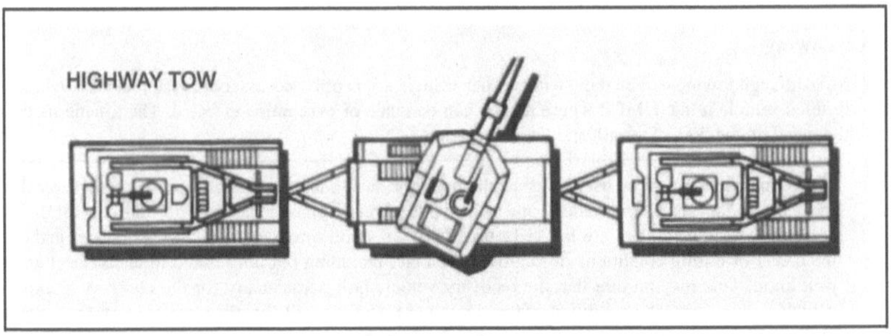

Figure 4-10. Tracked towing with holdback vehicle

Cross-Country Tow

4-31. Only in extreme cases, or as a backup, will crossed tow cables be used as a method of towing cross-country (figure 4-11). Use crossed tow cables between the recovery vehicle and the disabled vehicle when towing similar vehicles. A holdback vehicle is required so the towed vehicle will not overrun the recovery vehicle.

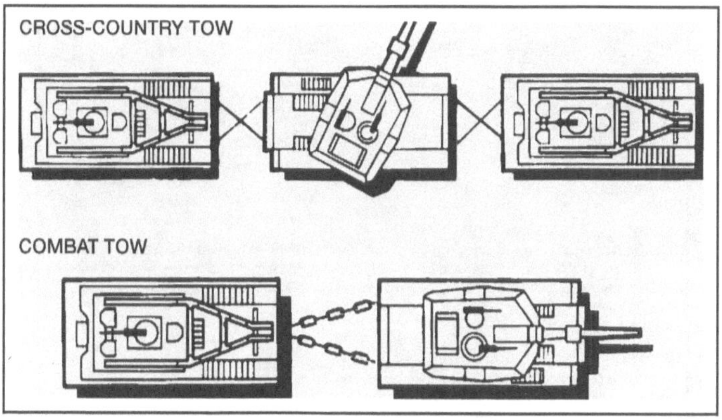

Figure 4-11. Tracked towing

Combat Tow

4-32. Use a combat tow to make a towing connection under small-arms fire to provide the least possible exposure of personnel (figure 4-11).

- Attach the lifting V-chain to the recovery vehicle's tow pintle before moving it to the disabled vehicle.
- Move the recovery vehicle into the danger area.
- Back it up until contact is made with the front of the disabled vehicle. (If possible, a crewmember in the disabled vehicle can connect the V-chain legs to the front tow hooks of the disabled vehicle.)
- Move out the recovery vehicle, towing the disabled vehicle.

Note. METT-TC/METT-T is a key factor in determining combat towing distances. Do not use combat tow for distances greater than one-fourth of a mile during peacetime operations. If conditions permit, change the towing procedure after that distance.

TOW BAR HANDLING

4-33. Before attempting to tow a disabled vehicle, be familiar with the location, features, and operation of all components of the tow bar. (Some tow bars have operator's instruction decals mounted on them.) Ensure the proper tow bar is used for the equipment being towed. Tow bars can be used to tow any vehicle up to the gross weight of the tow bar's towing capacity.

4-34. Before attaching a tow bar to a disabled vehicle, chock the wheels/tracks (figure 4-12) and/or set the emergency brake. After attaching the tow bar to a disabled vehicle, remove the chocks and/or release the emergency brake before moving.

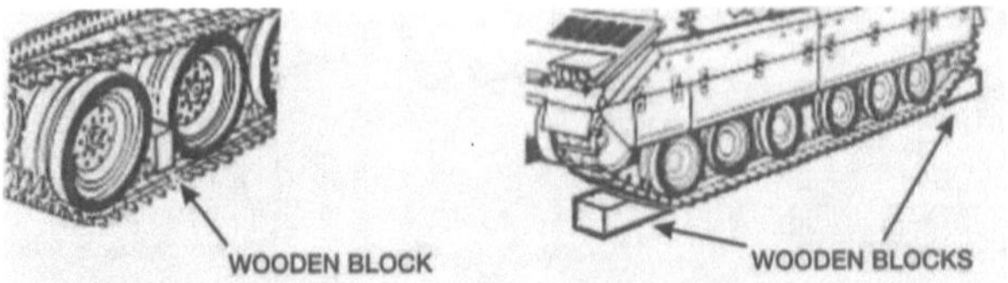

HOOKING UP TO A CARRIER:
Before hooking up the tow bar or disconnecting the drive between the differential and final drive, chock the vehicle with blocks so that it cannot move. (Place a block of wood or other suitable object between the track guides and the two sets of road wheels or one in front of and one in the rear of the track.) Make sure the object extends the full width of both road wheels.

UNHOOKING FROM A CARRIER:
Before unhooking the tow bar or hooking up the drive between the differential and final drive, chock the vehicle with blocks so that it cannot move. (Place a block of wood or other suitable object between the track guides and the two sets of road wheels or one in front of and one in the rear of the track.) Make sure the object extends the full width of both road wheels.

CAUTION
Failure to remove the blocks could result in damage to the vehicle.

Figure 4-12. Chocking/blocking tracked vehicles

4-35. Refer to the disabled vehicle's TM for proper towing procedures (for example, automatic versus standard transmission). Ensure that the proper pin assemblies are in the clevis holes, and always be sure the quick-release pins (which snap automatically) are properly secured.

Towing Principle

A key towing principle is to keep as much weight as possible on the front wheels of the towing vehicle. Do this by keeping the towed vehicle as close as possible to the towing vehicle without causing damage to either vehicle.

Note. For M936 and M1089 wreckers, raising the boom slightly (one of three holes), not extending the boom, or only extending it as little as possible and keeping the tow bar as short as possible, will all help keep the weight on the front axle of the towing vehicle. This only applies to the M936 and M1089 wreckers because they lose weight and steering ability as the load on the boom or lift device increases.

Note. Use extreme care when lift-towing or flat-towing disabled vehicles. Check the disabled vehicle's technical manual for vehicle preparation, precautions that must be taken, and maximum vehicle speed.

TOWING OPERATIONS ON GRADES

4-36. Towing a disabled vehicle is never easy, but towing up or down a grade can be even more difficult and dangerous. While towing a disabled vehicle, do not attempt to negotiate a grade (either up or down) greater than 25 percent unless there is express permission from the commander.

CAUTION

Under NO circumstances will a slope that is greater than 30 percent ever be negotiated while towing a vehicle.

4-37. To know which grades to avoid, an operator must know how grades are classified. Grades are defined in terms of percent or the amount of a grade's vertical height (rise) over its horizontal length (run). If a road gains 25 feet of height over 100 feet of length, it is classified as a 25 percent grade (figure 4-13).

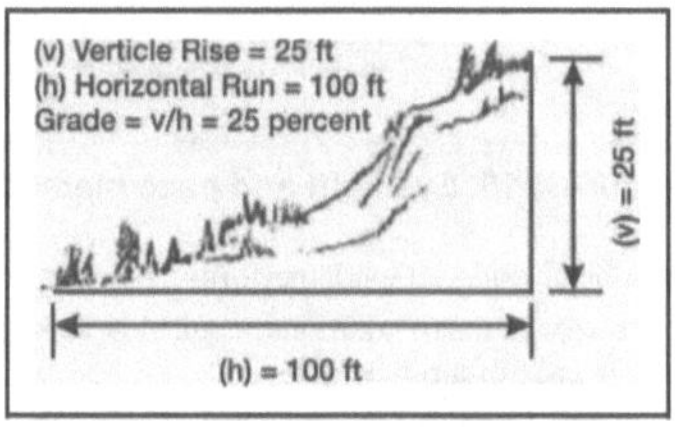

Figure 4-13. Classifying a grade

4-38. The best way to classify a grade is with a surveying level (figure 4-14), which is a BII on the M88A2. The operator stands at the top (or bottom) of the hill and chooses a point as close as possible to the bottom (or top) of the hill where he will be traveling. The operator then looks through the sight of the level at the point he has chosen and turns the level knob until he sees the level bubble centered between the witness marks. Then the operator reads the percent grade on the indicator.

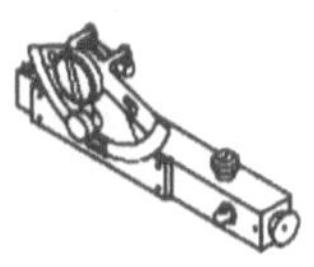

Figure 4-14. Surveying level

4-39. An improvised or field expedient method uses a small level, a 10-inch piece of flat wood, and a ruler.

- Lay the piece of wood on the steepest part of the grade, with the length of wood running up and down hill.
- Put the level on the piece of wood and start to raise the downhill side of the wood up, until the bubble in the level is between the witness marks.
- Measure the distance between the road and the bottom of the wood. If it is 3 inches, there is a 30 percent slope; if it is 2.5 inches, there is a 25 percent slope; and so forth.

4-40. Another method is the eyesight and pace method (figure 4-15). The Soldier needs to know his height and the length of his stride. If a Soldier is 6 feet tall and his step is 2 feet long—

- He stands at the bottom of the hill and picks a spot on the hill that is the same height as his head.
- He then walks to that spot, counting his steps.

- Once he reaches that spot, he multiples his steps by his stride (2 feet) and then divides his height (6 feet) by that number multiplied by 100 and adds 1.

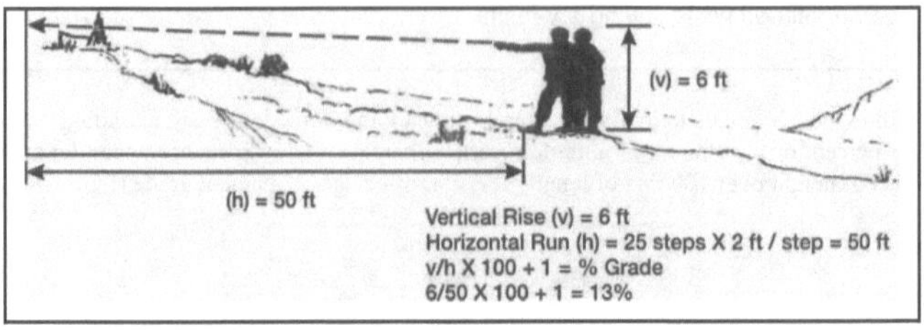

Figure 4-15. Eyesight and pace method

4-41. The following items must be considered while performing terrain analysis:

- Trails/grades with sharp curves mean additional control is needed during ascending and descending. There is no safety zone in case of a runaway load.
- Dry and dusty soil or wet and muddy soil can cause a loss in traction. Pay close attention to the soil conditions that may alter as weather conditions change.

Note. Inclement weather (rain, snow, ice) will naturally affect the road conditions, making loss of traction more probable.

4-42. If the operator has to shift into first gear to climb a grade, there is a good chance that it is too steep to descend with a towed load. Measure the downhill grade before attempting to descend any hill that required first gear to climb. Other options may include—

- Using a braking or holdback vehicle behind the towed load.
- Winching the disabled vehicle downhill, if possible.

4-43. If a way can be found to detour the steep grades, use it. If not, notify the commander. Inform him of the grade percentage of the road, weather visibility, and road conditions (wet, dry, muddy, pavement). The recovery vehicle driver's experience and the type of load being towed will play an important role in the commander's decision. If the driver does not feel confident in negotiating the grade, he must inform the commander. The best course of action may be to get the most experienced wrecker/recovery vehicle operator on the site to handle the mission.

4-44. In summary, ensure a good route reconnaissance is conducted on the way to the disabled vehicle's site. When possible, avoid all hills or roads with a grade of 25 percent or greater while towing a load. If not, notify your commander and take proper precautions. Ensure that no one rides in a towed vehicle.

Recovery and Towing with the Fifth Wheel Towing Device

4-45. The fifth wheel towing device (FWTD) is a heavy-duty, under lift towing device that uses the fifth wheel coupling as a pivotal connection between the pulling tractor and the truck in tow. (figure 4-16) The weight of the towed vehicle is transferred evenly to all axles of the towing tractor. The front axle of the towing tractor actually gains weight as the towed vehicle is lifted.

4-46. The 30,000 pounds lift capacity allows the FWTD to be used in recovering mired vehicles. The boom assemblies can be extended under the mired vehicle and the mast raised to allow for rigging attachments. This helps reduce the mired condition of a vehicle while transferring some of the weight to the recovery tractor.

4-47. The FWTD is equipped with an onboard winch to aide in coupling the disabled vehicle to the FWTD. Once the disabled vehicle is coupled to the FWTD, the system functions are similar to a semi-trailer. The FWTD transports wheeled vehicles on highways, unimproved roads (graded level), and cross-country.

CAUTION

The winch on the FWTD is not designed to recover vehicles in mired conditions. It is to be used only to assist in loading a disabled vehicle.

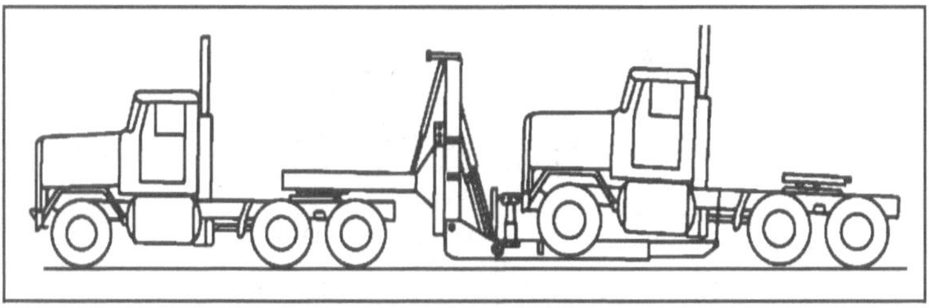

Figure 4-16. Fifth wheel towing device

4-48. Remember these important safety considerations when using the FWTD to transport or recover disabled vehicles:

- Towing a single vehicle with nonfunctioning brakes must be limited to not more than 25 mph on the highway and 15 mph off road.
- Stopping distances greatly increase when the towed vehicle has nonfunctioning brakes.
- Towing of vehicle combinations (tractor with trailer) with nonfunctioning brakes is prohibited.
- Visibility from the prime mover is significantly reduced when backing, whether the FWTD is loaded or not.
- All wheels remaining on the ground of the towed vehicle should be serviceable to increase system stability and reduce the risk of further damage.
- Never stand between the prime mover and FWTD when the prime mover is being backed up to the FWTD. Serious injury or death may result.
- Proper procedures must be followed and extreme caution used when backing to prevent damage to equipment and injury or death to personnel. See operator's manual for additional cautions for the FWTD.

4-49. Prime movers for the FWTD are listed in table 4-1.

Table 4-1. Prime movers for FWTD

Prime Mover	*Authorized Towed Vehicles*	*Gross Combination Weight Rating (GCWR)*
*M915, A1, A2, A3, A4	M915s, M818, M931 and M932 Models, FMTV Series, PLS Series	105,000 lb 47,641 kg
M916, A1, A2, A3	M915s, M916s, M818, M931 and M932 Models, FMTV Series, PLS Series	120,000 lb 54,446 kg
M920	M915s, M916s, M920, M931 and M932 Models, FMTV Series, PLS Series	120,000 lb 54,446 kg
M931	M911, M915s, M916s, M920, M818, M931 and M932 Models, FMTV Series, PLS Series	59,589 lb 27,036 kg
M932	M911, M915s, M916s, M920, M818, M931 and M932 Models, FMTV Series, PLS Series	59,742 lb 27,106 kg
M818	M818, M931 and M932 Models, M915, FMTV Series	57,665 lb 26,164 kg
M1088	M911, M915s, M916s, M920, M818, M931 and M932 Models, FMTV Series, PLS Series	80,775 lb 36,649 kg
M911	M911, M915s, M916s, M920, M818, M931 and M932 Models, FMTV Series, PLS Series	120,000 lb 54,446 kg
M983	M911, M915s, M916s, M920, M818, M931 and M932 Models, M983, FMTV Series, PLS Series	100,000 lb 45,372
* Note. M1074 and M1075 (PLS series) can be transported only without a payload.		

Allied Kinetic Energy Recovery Rope

4-50. The AKERR kit consists of a multi-strand, woven nylon rope, two large shackles, four small shackles, and a canvas bag. The concept of operations behind the AKERR is that the stretch of the rope and its subsequent attempt to return to its normal length will provide a sudden snatch effect. This effect provides additional recovery force.

4-51. The AKERR is restricted to recovering full tracked, full armored vehicles. Only vehicles using the proper attaching equipment (as listed in TM 9-4020-200-10, page 1-3) may participate in AKERR recoveries as either the recovering vehicle or the vehicle to be recovered.

4-52. Any full track, full armored vehicle can be recovered. However, maximum effectiveness will be achieved when the vehicle being recovered is within the same general weight range or lighter than the recovering vehicles. Vehicles that are authorized by the technical manual to be the recovering vehicle are listed in table 4-2.

Table 4-2. Vehicle weights	
M11OA2	62,500 lb (28,350 kg)
M109A2/M109A3	55,000 lb (24,948 kg)
M2/M2A1-M3/M3Al	43,600 lb (19,777 kg)
M992	57,500 lb (26,082 kg)
M578	54,000 lb (24,494 kg)
M551/M551A1(with 1 1/4 inch tow cables)	36,000 lb (16,330 kg)
Note. Only M5511M551Al vehicles with 1 1/4 inch diameter tow cables are authorized to be recovering vehicles. Those with 3/4-inch tow cables are not authorized to be recovering vehicles.	

CAUTION

The AKERR may fail during use. If metal attachments fail or the rope breaks while the rope is under tension, personnel may be injured or killed. All personnel must either remain inside vehicles with hatches closed or stay to the side at least 100-125 feet away from the vehicles during recovery.

4-53. To use the AKERR, the towing vehicle travels in reverse as closely as possible to the mired vehicle. The rope is connected and snaked to allow tangle-free deployment. For situations where it is not possible to get close to the mired vehicle, extension cables may be used. (Rigging for individual vehicles is shown in TM 9-4020-200-10.) The AKERR must be connected directly to the vehicle.

Note. The gears are engaged in the mired vehicle to allow for travel in the same direction.

4-54. On a prearranged signal, the recovery vehicle will accelerate at maximum speed. At the same time, the mired vehicle (if possible) will accelerate to assist the recovery effort. Maximum acceleration is continued until the recovery vehicle is stopped or the mired vehicle is recovered (figure 4-17).

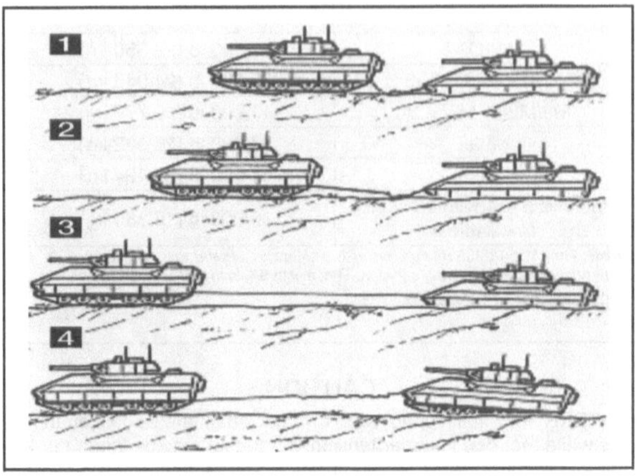

Figure 4-17. Recovery with AKERR

4-55. After acceleration, the recovering vehicle will slow or halt; its kinetic energy is converted into potential energy as the rope stretches. The rope transfers the energy into the mired vehicle. After a slight pause, the mired vehicle rises free. If the vehicle is not freed by the first attempt, the process should be repeated. After recovery is complete, remove the AKERR, wash it with low-pressure water, and allow it to dry before stowage.

SPECIAL RECOVERY SITUATIONS

NOSED TRUCK

4-56. The recovery of a nosed truck using a wrecker may require only a towing operation. However, some situations may require using all three of the wrecker's capabilities (winching, lifting, and towing) to complete the recovery.

4-57. Figure 4-18 shows an example of a 2½-ton cargo truck that is nosed off of a narrow road and mechanically disabled. Although the apparent fleet angle of the winch cable in the figure is greater than 1½ degrees, the wrecker winch has a level winding device which offsets the difference. (Not all vehicles with winches have this device.) When possible, position the wrecker at the least possible fleet angle and on the most solid surface to improve stability.

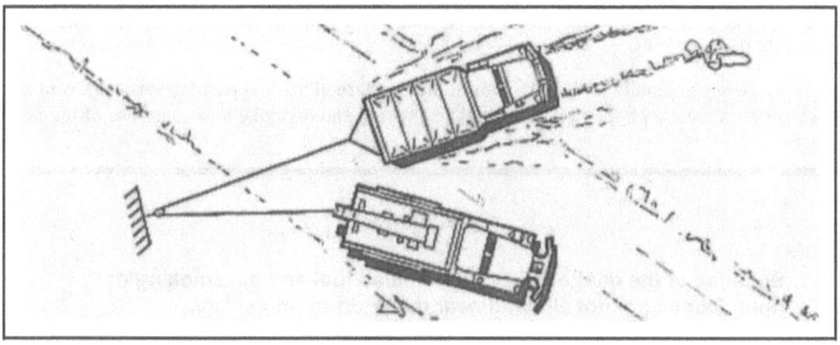

Figure 4-18. Recovery of a nosed cargo truck

4-58. To perform the recovery—

- Position the wrecker truck on the road so the front end of the nosed truck, when pulled back up on the road, will be in line with the rear of the wrecker truck.
- Make a change of direction pull, using the wrecker's rear winch to pull the truck onto the road.
- Lift the front of the truck with the wrecker's outriggers in place and turn the crane to place the truck directly behind the wrecker truck to prepare for towing.

OVERTURNED TRUCK

4-59. To upright an overturned truck (figure 4-19) using the wrecker, a sling method of attachment must be used because a pulling force applied to only one point of the frame may result in a bent frame.

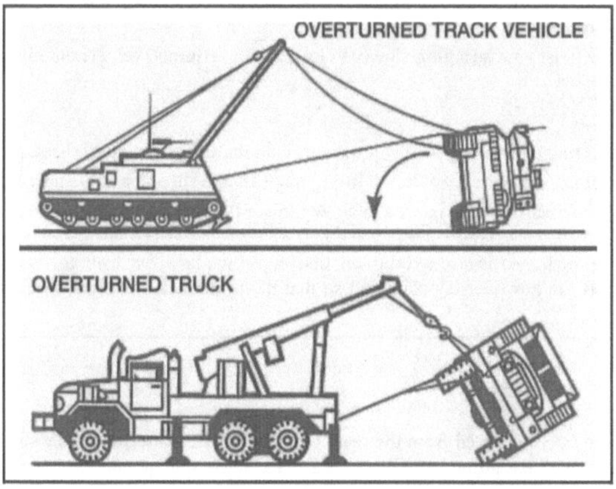

Figure 4-19. Recovery of overturned vehicles

4-60. A sling attachment is made of two utility chains. The sling ends are attached to the front and rear lifting shackles on the high side of the overturned truck. Then the winch cable is attached to the center of the sling.

4-61. A holding effort will be required to prevent the overturned vehicle from crashing onto its wheels. (The holding force could be another vehicle, the wrecker boom, or a rope block and tackle with manpower.) The attachment for the holding force is a holding sling attached to the same points on the overturned truck as the pulling sling. The holding sling is then attached to the holding force with cable, rope, or chain—making sure the holding force is attached to the center of the sling. If a holding vehicle is not available, use the wrecker boom to hold the load.

4-62. Apply power gradually to the winch until the overturned truck is past the vertical position. Then, lower the truck on its wheels with the hoist winch. This method should make maximum use of the boom jacks and outriggers.

WARNING

Because of the danger of igniting spilled fuel and oil, smoking or open flames are not allowed near the overturned vehicle.

Overturned Tracked Vehicle

4-63. To upright an overturned tracked vehicle with a recovery vehicle, position the recovery vehicle so it is facing the bottom of the overturned vehicle. It should be at a distance equal to the width of the overturned vehicle, plus 2 feet for safety.

4-64. For the holding source of power, rig the boom with its maximum MA rigging and attach its hoist block to two tow cables to form a sling. Pass the opposite ends of the sling under the track. Attach them to the front and rear tow hooks on the high side of the overturned vehicle.

4-65. For the uprighting source of power—

- Use a utility chain to attach the main winch cable to the center road-wheel arm support housing on the high side.
- Apply power to the main winch until the vehicle pulls past its point of balance and is supported by the hoist rigging.
- Lower the hoist winch rigging slowly to lower the overturned vehicle onto its suspension system.

Howitzer

4-66. When uprighting a howitzer, follow procedures similar to uprighting a wheeled vehicle.

- Attach the tow cable or chain to the lifting loops on the shoulder of the howitzer.
- Ensure a holdback vehicle is used to slowly lower the howitzer to the ground.

Note. If the prime mover is also overturned, disconnect the howitzer from the vehicle and upright the vehicle first. If the howitzer is positioned so that the prime mover cannot be uprighted, upright the howitzer first.

Forklifts

4-67. Forklifts can only be towed from the rear. Towing forklifts forward poses a serious overturn risk if the forks make contact with the ground and dig in. Refer to the equipment operator's manual for specific towing instructions. To upright an overturned or mired forklift, use the overturned vehicle recovery procedure or the mired-vehicle procedure.

Armored Vehicle-Launched Bridge

4-68. When recovering the armored vehicle-launched bridge (AVLB), the bridge has to be removed by another AVLB using a hydraulic slave procedure. (The bridge cannot be removed by the M88 hydraulic system because the couplings differ in design.) Once the AVLB is removed from the prime mover, refer to the operator's manual for towing and hookup procedures.

COMBAT ENGINEER VEHICLE

4-69. Make sure the road wheels are chocked before disconnecting the final drives. When a tow bar or cables are used, a second vehicle is required when descending a grade of 20 degrees or more. A second vehicle is also required when the road conditions dictate. A combat engineer vehicle should be towed only from the rear and only when the blade has been removed from the front of the vehicle.

MINE PLOW AND MINE ROLLER

4-70. Vehicles with mine plows and mine rollers attached cannot be towed from the front until the mine rollers or mine plows have been removed. These vehicles can be towed from the rear provided the terrain and situation permit. Once it is determined how the vehicle will be towed, refer to the operator's manual for towing and hookup procedures.

CRANE, WHEEL-MOUNTED

4-71. This vehicle can be towed but information on road conditions and possible restrictions along the route must be obtained.

- Use a vehicle with an air brake system capable of producing 120 psi in the system.
- Place the boom over the front—which is the most stable position for towing.
- If towing more than one-fourth of a mile, disconnect the propeller shafts from the front and rear axles.

Caution must be used when turning and traveling through towns. For information on safely using cranes and similar equipment, refer to TB 385-101.

ROAD GRADER

4-72. When towing the road grader for distances greater than half a mile, maintenance personnel must remove the tandem drive chains. If the distance is less than half a mile, it is not necessary to remove the tandem drive chains, but the speed must be kept below 5 miles per hour. When maintenance personnel are not available to remove the tandem drive chains and the distance is more than half a mile, trailers must be used to recover the road grader.

SCOOP LOADER

4-73. This vehicle should not be pushed or towed. A flatbed trailer must move this vehicle. In the event of an emergency where the scoop loader must be towed, the maximum distance the loader may be towed or pushed is half a mile—at low speed not exceeding 5 miles per hour. Refer to the operator's manual.

M9 ARMORED COMBAT EARTHMOVER

4-74. The M9 armored combat earthmover (ACE) must be towed from the rear and the final drives must be disconnected to prevent damage to the steering unit. When turning with the ACE in tow, turn in a wide arc to prevent undue strain on the suspension of the disabled vehicle and tow bar. Make sure the disabled vehicle is in the SPRUNG position. Refer to the operator's manual for additional towing information.

NOSED TRACKED VEHICLE

4-75. Various factors must be considered before recovering a tracked vehicle nosed in a deep trench or ravine. If the terrain behind the nosed vehicle is level, recover by towing. If the terrain is not suitable for towing, perform a winching operation (figure 4-20) as follows:

- Move the recovery vehicle to the opposite side of the trench or ravine (to the front of the nosed vehicle).
- Using the recovery vehicle's boom with its maximum MA rigging, attach its hoist block to the front lifting eyes on the nosed tank with a V-chain.

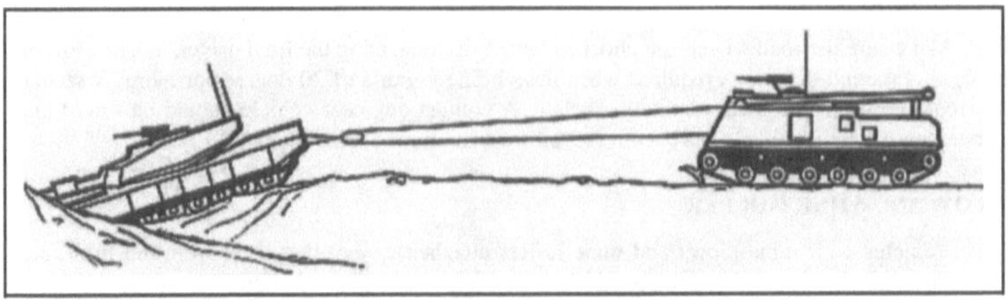

Figure 4-20. A recovery vehicle winching a nosed tracked vehicle

- Lift the vehicle to a horizontal position and pull it to the opposite side of the ditch where towing or winching can complete the recovery (figure 4-21).

Figure 4-21. Lifting operation

Note. Nosed vehicles may be recovered by towing, winching, or lifting operations. If the tow hooks are accessible on the nosed vehicle, use rigging to attach them to the winch. Recover the nosed vehicle with a combination of winching and hoisting. Control the weight and movement of the disabled vehicle during the entire recovery operation by coordinating the hoist winch and the main winch.

SELF-RECOVERY AND LIKE-VEHICLE RECOVERY

SOURCE OF EFFORT

4-76. The amount and type of equipment used as the source of effort during any recovery operation depends on the level of recovery. Drivers and crews should evaluate the situation and determine if the crew can recover the vehicle before calling on support from a higher level. During combat, it may be imperative that cargo reach its destination at a definite time, that the personnel or cargo be picked up at a given time, or that a combat vehicle be at a given place at a specific time.

4-77. Using like vehicles is usually the quickest method of recovery because they are readily available. Call for recovery support only when self-recovery or like-vehicle recovery techniques cannot support the recovery operation. A mired vehicle with no winch may be freed by using recovery expedient measures discussed previously.

4-78. Combat vehicles, which need fuel, ammunition, or repairs not related to mobility (for example, a fire control malfunction), can tow disabled vehicles to the refuel, rearm, or maintenance site. However, engaged combat vehicles should never be diverted for recovery. METT-TC/METT-T will be the determining factor for like-vehicle recovery.

4-79. Use like-wheeled vehicles as the source of effort to perform recovery by towing and winching. (For vehicles not equipped with lifting shackles, attach a tow chain to the main structural members.) Before towing or recovering a disabled vehicle, check the vehicle's TM to ensure all physical and safety features are considered (for example, automatic transmissions, fail-safe braking systems, and articulation). This must be done so the disabled vehicle is not damaged further.

RECOVERING A MIRED TRUCK

4-80. To recover a mired truck by towing with a like vehicle, use a tow chain, cable, or bar between the towing vehicle and the mired vehicle.

- Attach a tow chain, cable, or bar to one lifting shackle (both, if possible) of the mired vehicle and the tow pintle on the towing vehicle. If a greater working distance is required to enable the towing vehicle to get better traction, use the towing chains or other device from both vehicles.
- Apply power slowly to prevent placing an impact load on the towing device and lifting shackles. A chain, unlike a cable, will not stretch and can be broken easily by impact loading. If one towing vehicle cannot attain sufficient towing effort to overcome the resistance, use another towing vehicle in tandem with the first (figure 4-22).

WARNING

Rigging should include a chain or Y-sling attached to both lifting shackles whenever possible to minimize damage and create an even pull effort. The risk of tearing off lifting shackles, bending the rear cross member, or distorting the frame of a mired vehicle is even greater if a second vehicle is used in tandem.

Figure 4-22. Recovering a mired cargo truck in tandem

PROPER HOOKUP WITH A LIKE VEHICLE

4-81. To recover a mired cargo truck, use a truck of an equal or heavier vehicle class as an anchor vehicle (figure 4-23). Use the winch (if equipped) mounted on the mired vehicle to perform the winching operation. A mired 2½-ton cargo truck may be winched with either a 2½-ton or a 5-ton vehicle only under emergency conditions and a snatch block must be used in the rigging. All winch-equipped trucks are authorized a single sheave snatch block and one tow chain for rigging.

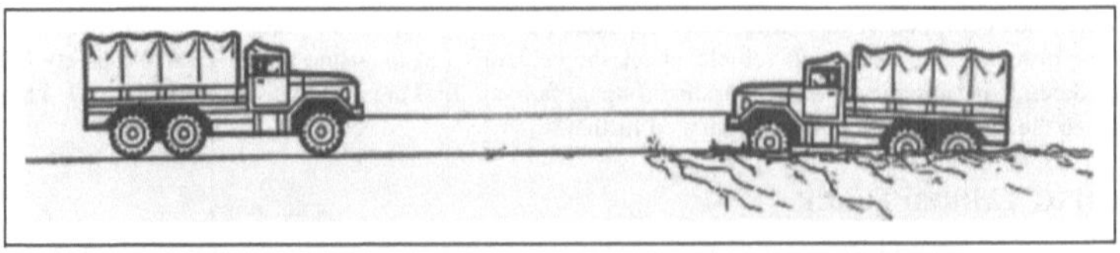

Figure 4-23. Winching with like or heavier class wheeled vehicle

4-82. Determine if the resistance of the mired truck is greater than the winch capacity. If it does not exceed the winch capability, MA is **not** required. Proceed as follows:

- Position the anchor vehicle in line with the mired vehicle so the correct fleet angle is obtained.
- Free-spool the winch cable from the drum.
- Attach a Y-sling or chain to both front lifting shackles of the anchor vehicle, and attach the winch cable clevis to the apex of the sling or the center of the chain. (See figure 3-7 for proper angles and dimensions for Y-slings.) The angle of the Y-sling must be less than 30 degrees to reduce strain on the lifting shackles.

4-83. If MA **is** required, proceed as follows:

- Attach a snatch block to the center of the chain or apex of the Y-sling and the winch cable routed through the snatch block back to the mired vehicle.
- Place the loop formed in the winch cable in the snatch block.
- Apply power to the winch to remove the slack from the cable.

If the anchor vehicle must be anchored by more than just its weight, place wheel blocks, chocks, or natural material in front of the anchor vehicle's front wheels. (See chapter 3).

SELF-RECOVERY

4-84. A winch-equipped, mired vehicle can perform self-recovery using an anchor (figure 4-24). Attach the snatch block to a suitable anchor, and attach the free end of the cable to a chain sling connected to both of the mired vehicle's front lifting shackles. A fixed block can be used to gain MA when performing a self-winching operation.

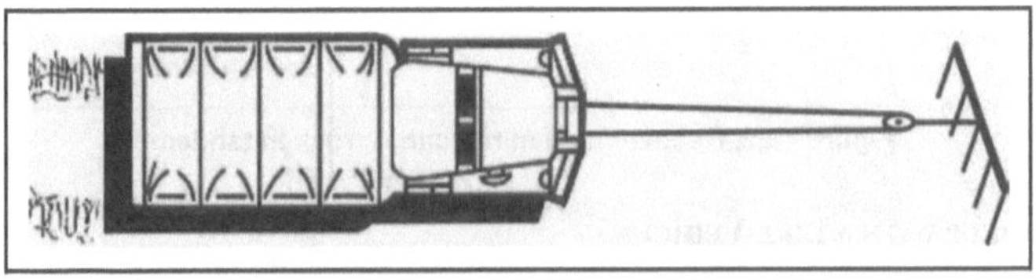

Figure 4-24. Self-recovery operation

USE OF LIKE-TYPE TRACKED VEHICLES FOR RECOVERY

4-85. The number of tracked vehicles required for a specific recovery depends on the resistance to be overcome, the type of disablement, and the terrain conditions. To rig for recovery, attach the tow cables to the tow hooks of both vehicles. All main battle tanks carry two tow cables. Light-tracked vehicles carry one tow cable.

4-86. When a vehicle with a main gun cannon tube is recovered or towed, rotate or elevate the gun tube. This prevents serious damage if the rigging fails or the towed vehicle rams the towing vehicle.

4-87. When using two tow cables between two vehicles, make sure the cables are crossed (figure 4-25). If a greater working distance between the pulling vehicle and the mired vehicle is required, join the tow cables together with tow hooks.

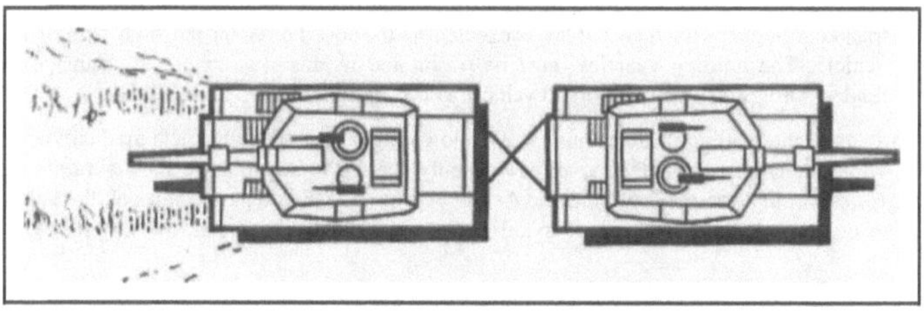

Figure 4-25. Recovery of a mired tank using one like vehicle

4-88. If two vehicles are required for an operation, one tow cable is enough because the strength of one tow cable is slightly greater than the pulling effort of the second pulling vehicle. However, use two tow cables, when available, to maintain alignment and equalize the pulling effort. When using two vehicles, turn the gun tube of the center vehicle to the side to prevent contact and possible damage.

RECOVERING A NOSED TRACKED VEHICLE WITH LIKE VEHICLES

4-89. As many as three like vehicles may be needed to recover a nosed-tracked vehicle (figure 4-26). This depends on the degree to which the vehicle is nosed and the terrain conditions on which the pulling vehicles must operate. In extreme instances, another resource may be required to lift the front of the nosed vehicle.

- Position the lifting vehicle to face the nosed vehicle.
- Connect the cables of the pulling vehicles in the same way as for recovering a mired vehicle.
- Apply power to all assisting vehicles at the same time. The front of the nosed vehicle will rise and move toward the rear.
- Slowly move the lifting vehicle forward.
- Support the vehicle until it is recovered.
- If any oil or fuel has spilled in the nosed vehicle, do not run the engine until the spill is cleaned.

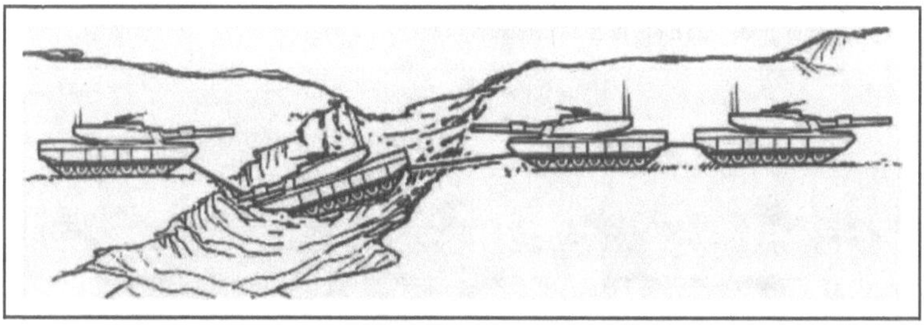

Figure 4-26. Recovering a nosed tracked vehicle with like vehicles

RECOVERING AN OVERTURNED TRACKED VEHICLE WITH LIKE VEHICLES

4-90. An overturned tracked vehicle can be uprighted by using three like vehicles (figure 4-27). Use one vehicle to pull the overturned vehicle upright. Use the other two vehicles to hold and retard the fall of the overturned vehicle so it does not crash down on its suspension system.

- Connect tow cables together in pairs to allow a safe working distance.
- Connect the cable used to upright the overturned vehicle to the nearest center road-wheel arm support housing on the upper side of the overturned vehicle. Never connect to any other part of the suspension system, turret, or the tie-down eyes.
- Position the two vehicles used for holding at a 30-degree to 45-degree angle from the overturned tracked vehicle, with their cables connected to the tow hooks on the high side of the overturned vehicle. The holding vehicles must be positioned in this way to prevent damage to the cables, fenders, or lights of the overturned vehicle as it is uprighted.

4-91. Drivers of the holding vehicles must shift to low range. The pulling vehicle gradually applies power in reverse, while the holding vehicles move forward only enough to keep their cables taut until the overturned vehicle passes through the point of balance. As the overturned vehicle passes through the balance point, the holding vehicles move forward slowly, supporting the overturned vehicle and lowering it onto its suspension system.

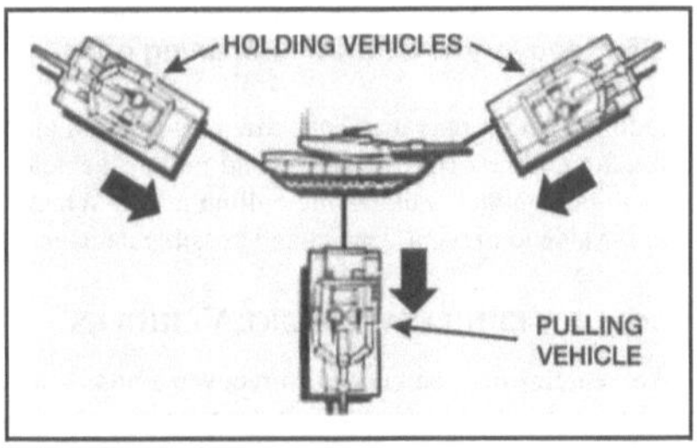

Figure 4-27. Recovering an overturned tracked vehicle with like vehicles

Towing Disabled Tracked Vehicles

4-92. Tow a disabled tracked vehicle with a like vehicle of the same weight class or heavier weight class with a tow bar or two tow cables (figure 4-28). When using a tow bar on vehicles lighter than the 70-ton class, no holdback vehicle is required, unless the terrain interferes. A holdback vehicle will be used when—

- Tow cables are used.
- The towed vehicle is heavier than 70 tons.
- Terrain grades are more than 20 percent.

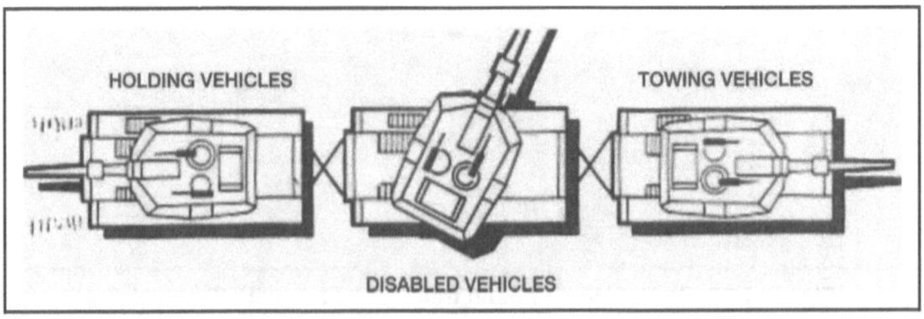

Figure 4-28. Towing a disabled tracked vehicle

4-93. Cross the tow cables to keep them from being tangled with the tracks. When towing tracked vehicles with only one track, there will be a difference in resistance and steering capability between a complete track on one side and road wheels on the other side. As a result, the towed vehicle will pull in the direction of the side lacking the track. Compensate for this pull by properly attaching the towing cables. Check the TM pertaining to the towed vehicle to determine the necessary preparations and precautions to be used to prevent further damage. Never exceed the towing speed stated in the TM.

4-94. If the disabled vehicle has defective brakes or its universal joints are disconnected, use another similar vehicle for holding (figure 4-29). Use crossed cables between the holding and disabled vehicles when available. With vehicles that are issued only one tow cable, the vehicles will be connected with crossed cables between the towing vehicle and the disabled vehicle. Connect a single tow cable between the disabled vehicle and the holding vehicle.

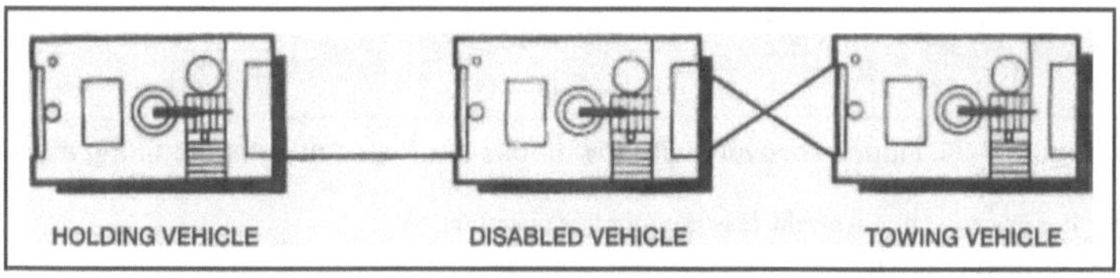

Figure 4-29. Towing with vehicles issued one tow cable

Note. Use crossed tow cables between the holding and disabled vehicles when enough cables are available.

MARINE RECOVERY

4-95. Many vehicles can now swim or ford. Some of these vehicles will fail while waterborne and will need recovery. Situations may be as simple as stalled, floating vehicles or as complex as submerged vehicles. The same methods of recovery apply to these situations but with a few unique considerations.

4-96. In the case of floating vehicles, swiftly moving current can carry the vehicle and crew downstream. Water safety must be stressed to both vehicle and recovery crews engaged in these operations. Current and stream bottom conditions interact effectively to bury a vehicle, thereby increasing resistance. When operating on beaches or rivers with soft bottoms, time is critical. Recover the vehicle as quickly and safely as possible.

Fording Vehicle

4-97. Vehicles become mired, nosed, and overturned during fording operations just as they do during land operations. As a result, estimate resistance in the same way by considering vehicle weight and type of disablement.

Swimming Vehicles

4-98. A mechanically disabled swimming vehicle offers little resistance while floating. Compared with its rolling resistance on land, it can be recovered with little effort. The same rigging is applied in floating vehicle recovery as in land recovery. The only exception is that the attachments are made to the lifting eyes instead of the tow lugs. This prevents the crew from having to work in the water (figure 4-30).

Figure 4-30. Marine recovery with tow hooks and ropes attached to lifting eyes

4-99. For example, if a like vehicle is used for this operation—

- Attach its tow hooks to the lifting eyes before entering the water.
- Cross the towropes and attach to the lifting eyes until the disabled vehicle is towed to shore. (Using cables will prevent the quick disconnect of the towing vehicle if the towed vehicle begins to submerge. The axe that must be with the rope is used to cut the rope in an emergency.)
- Once the vehicle is close to the shore and the tow lugs are exposed, move the tow cables to the tow lugs on both vehicles to pull the disabled vehicle ashore (figure 4-31).

Figure 4-31. Recovery to shore with tow hooks and cables attached to tow lugs

Submerged Vehicles

4-100. If a vehicle is flooded and submerged, determine the resistance on the river bottom in the same way as on land. Consider the weight of the vehicle, the cargo, and the river bottom, which may be sand, gravel, or mud. In addition, when pulling flooded vehicles from water to land, consider the weight of the water when determining the resistance. Water weight is estimated to be equal to the vehicle's weight. For example, a tracked vehicle weighing 52,000 pounds sank. The vehicle is mired in the riverbed (mud) to fender depth. The effort required to retrieve it is 156,000 pounds (2 x 52,000 - pound mire factor + 52,000 - pound water weight).

4-101. The first problem in underwater recovery is locating the disabled vehicle in deep water. It may be easier to use dragging devices to locate the vehicle. Divers can then determine the location of the vehicle's rigging and mark the location of the vehicle using lines and floats. Special purpose vehicles, such as wrecker trucks and recovery vehicles, are readily adaptable to recovery operations on submerged vehicles. In most situations, the winch cables of the recovery vehicles are long enough to allow winching operations from water to land.

WATER OPERATIONS

4-102. Most vehicles currently in the inventory of the U.S. military have either a swim or fording capability. Vehicles involved in fording, swimming operations sometimes become disabled from mechanical or mobility malfunctions. A vehicle that has swim capability will usually remain afloat even if the main engine fails. This is possible because amphibious vehicles are usually equipped with an auxiliary engine and bilge pumps.

4-103. If a vehicle is completely disabled during water operations, it must have power restored using BDAR or any other means available. Amphibious vehicles are at the mercy of the surf or river current when power is lost. If left afloat without power, vehicles are at risk of sinking, causing further damage to the vehicle and serious water contamination. If sinking does occur, all practicable efforts should be made to avoid environmental contamination. Contamination over 1 gallon should be reported through the chain of command. Should a vehicle become submerged (out of sight), qualified scuba personnel should be called to assist in locating and rigging the vehicle for recovery.

Resistance in Water

4-104. Water resistance occurs when submerged vehicles are pulled from water to land. Water resistance is estimated as additional resistance equal to the vehicle weight. Therefore, a vehicle weighing 25 tons (including cargo) would require 50 tons of effort to winch it from the water. In the same situation, resistance would increase if the vehicle went down in the surf and the sand was partially covering the vehicle. Vehicles completely submerged, even for a short period, will usually be in a mired condition from sand, if in the ocean, or mud, if in a river. If in doubt, rig for the greater resistance.

4-105. Whether the vehicle is upright or overturned will also be a factor in determining the total resistance. Again, qualified divers should be used to locate and rig a vehicle for recovery. They will also be able to recommend direction of recovery, depending on obstacles. Following are some examples of resistance encountered when recovering floating-type vehicles:

- Amphibious vehicle afloat, minimal – 1/64th of vehicle weight.
- Amphibious vehicles completely submerged – equal to the weight of the vehicle. If the vehicle is mired on a river or ocean bottom, calculate the additional resistance the same as for land mire.
- Amphibious vehicles completely submerged and filled with water, the submerged vehicle weight is the vehicle weight times two.
- Fording-type vehicles that have become disabled must also be considered for weight of water but only an additional 1/8th of the vehicle weight; that is, a 70-ton tank would be calculated to weigh approximately 79 tons plus any mire encountered. The mire factor in this case is figured using 79 tons.

4-106. During underwater recovery operations, air bags can be placed inside the submerged vehicle and inflated to provide buoyancy and decrease resistance. To employ air bags in this type of situation, qualified divers are recommended. The air bags or 55-gallon drums need to be placed inside the vehicle in a location where they will not escape the vehicle or cause additional damage. Once the air bags are in position, inflate to the recommended capacity.

Methods Of Rigging

4-107. The rigging methods for underwater recovery are normally restricted to the manpower and/or lead methods. Towing from water is recommended only if the disabled vehicle is located in very shallow water. The method of rigging depends on the distance from the disabled vehicle; the type of disabled vehicle; the type of recovery vehicle available; the equipment available (floats, air bags, tackle); and the condition of the disabled vehicle.

Lead Method

4-108. The lead method of rigging is performed the same in water as on land. If in deep water, a boat or an amphibious vehicle can transport tackle to the disabled vehicle. If the water is shallow, the tackle can be manually carried to the disabled vehicle.

Manpower Method

4-109. The manpower method is much the same regardless of whether on water or land. However, flotation devices can be attached to cable every few feet or to snatch blocks and other tackle to aid in getting the recovery equipment to the disabled vehicle.

Note. Underwater recovery is usually limited to the manpower or lead methods.

Chapter 5
Field Expedient Recovery

An expedient measure is any method by which a task is done using on-hand materials. For example, vehicles may be required to operate in remote areas where assistance in recovery operations is not readily available. Under these conditions, the operator or crew must attempt self-recovery by using methods similar to those described previously in this manual.

RECOVERY EXPEDIENTS

PRY BAR

5-1. A pole can be used to pry a lightweight truck out of a ditch (figure 5-1) by—

- Using the pole to lift the front end of the truck.
- Applying power to the truck while in reverse gear.

Figure 5-1. Pry bar recovery

SUBSTITUTE JACKS TO REMOVE FRONT AND REAR WHEELS

5-2. To raise the front wheel of a cargo truck—

- Locate a timber (approximately 5 feet long) to use as a pry bar.
- Place the bottom of the timber in a shallow hole.
- Secure the timber to the front bumper at an angle with a chain or rope.
- Move the vehicle forward until the timber is in a vertical position and the wheel clears the ground. (This "substitute jack" is shown in figure 5-2.)
- Set the brakes and chock the wheels.

Figure 5-2. Substitute jack (front wheels)

> **WARNING**
>
> **This method cannot be used with vehicles with aluminum front bumpers, such as the USMC MK-23. If this method is used, it will damage the front end of the vehicle and possibly cause injury to personnel.**

5-3. When an **outside** rear dual tire is flat and a jack is not available, run the **inside** dual wheel up on a small log or rock (figure 5-3). This takes the weight from the outside wheel, which can then be removed for replacement.

5-4. Another substitute jack is a piece of timber longer than the distance from the axle to the ground (figure 5-3).

- Place one end of the timber against the axle at an angle and the other end in a shallow hole.
- Drive the vehicle forward against the angled timber, which will cause the timber to stand straight and lift the axle off the ground.
- Set the brakes and block the vehicle securely.

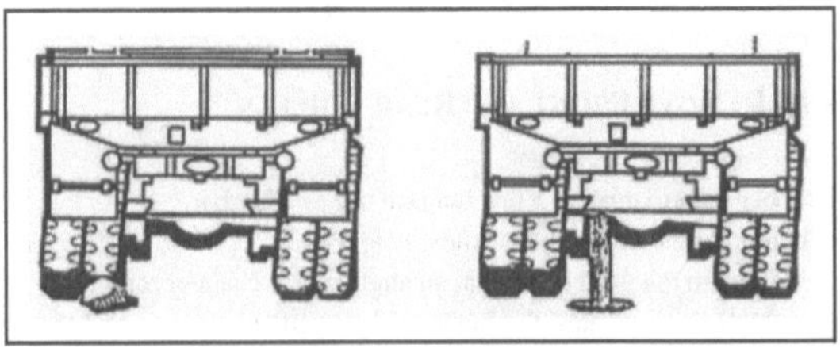

Figure 5-3. Substitute jack (rear or tandem wheels)

Dual Wheel Winching

5-5. With dual-wheel vehicles (not equipped with a winch), use the rear wheels to aid in recovery (figure 5-4).

- Fasten one end of the rope to the wheel hub and anchor the other end.
- Run the end of the rope fastened to the wheel's hub between the duals and through one of the holes in the wheel disk. (Do not place the rope through a hole in the wheel disk where the valve stem is located.)
- Tie a knot in the end of the rope and slip it over the hub.

- Tie a second rope in the same way to the dual wheels on the other end of the axle.
- Place the vehicle in reverse gear. The ropes will wind between the two duals and cause the vehicle to move rearward.

Figure 5-4. Wheels used as winches

USING AN A-FRAME

5-6. Frequently, a truck will become nosed in a shallow hole, narrow ditch, or munitions crater where both lifting and pulling forces are required to make the recovery. If the nosed wheeled vehicle is equipped with a winch, use the winch to supply the pulling force and an A-frame for the lifting force (figure 5-5).

Figure 5-5. Girth hitch and using an A-frame in recovery operations

> **CAUTION**
>
> This method cannot be used on vehicles with aluminum front bumpers such as the Marine Corps MK-23 series trucks.

5-7. To build an A-frame, use two long poles.

- Lash the poles together at the top using a figure eight or girth hitch.

- Place the lower end of the poles in the ground 10 to 12 inches deep to prevent them from sliding when power is applied.
- Lay the upper end of the A-frame against the front of the vehicle. Attach the A-frame to the vehicle.

5-8. A nosed wheeled vehicle's winch and an A-frame can supply the lifting and pulling forces needed to recover the vehicle.

- Rig the winch cable for a 2-to-1 MA.
- Secure the end of the cable to the apex of the A-frame. As the winch is wound in, the A-frame lifts and pulls the truck forward.
- Suspend the truck over the ditch or hole.
- Fill the ditch with rocks, dirt, or other material so the vehicle can be driven forward or backward.

CAUTION

Do not wrap chains around the bumper or vehicle frame. Attach chains to the vehicle by lifting shackles only.

REMOBILIZING TRACKED VEHICLES

ANCHORING TRACKS

5-9. Vehicles often become bellied (high-centered) on stumps, rocks, dry ridges, or mire. When this occurs, they become immobilized due to lack of traction.

5-10. To recover a bellied vehicle, obtain a log long enough to span the width of the vehicle and of sufficient diameter to support the vehicle weight (figure 5-6).

- Place the log against both tracks.
- Place a tow cable so that one end of the cable goes under the log and through the tracks from the inside.
- Place the other end of the tow cable underneath the log and connect the ends of the cable together with a tow hook on the outside of the track to make disconnecting easier.

Figure 5-6. A log used to anchor tracks

5-11. Follow the same procedure to attach the log to the track on the opposite side of the vehicle. Take up the slack in the tow cable by gradually applying power to the tracks. This pulls the log underneath the tracks until it comes into contact with the obstacle, thereby anchoring the tracks and causing the vehicle to move.

CAUTION

To prevent damage to the fenders and tow cables, stop the vehicle before the log reaches the fenders or the cable reaches the drive sprocket.

5-12. For a bellied disablement (other than mire), anchor the tracks by using two tow cables (figure 5-7). Connect the tow cables together with a tow hook and attach the cables to both tracks by passing the ends of the cables through the tracks from the outside and attaching them to the standing parts of the cables with tow hooks. When power is applied to the tracks, the cable will contact the obstacle and anchor the tracks.

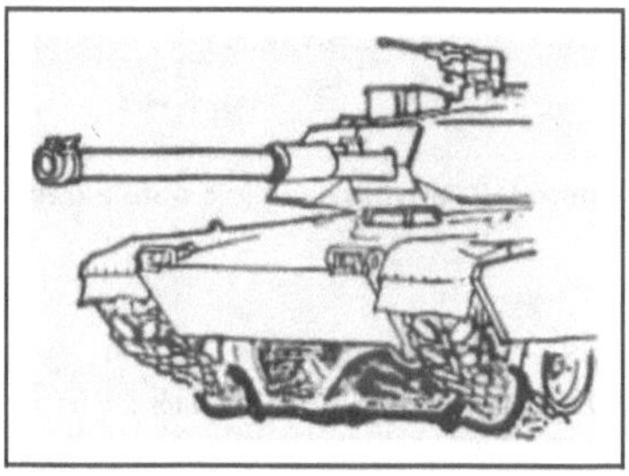

Figure 5-7. Cables used to anchor tracks

Moving a Vehicle with Both Tracks Broken

5-13. When both tracks are thrown, they may need to be separated before the vehicle can be moved to remount the tracks (figure 5-8).

- Break one track and attach a cable from the drive sprocket hub to an anchor. This will support the vehicle so that the other track can be separated.
- Chock the vehicle to keep it from rolling out of control.
- Apply engine and steering power to the drive sprocket attached to the cable. When this is done, the vehicle will move by the winching action of the drive sprocket hub.

Figure 5-8. Moving a vehicle with both tracks broken

MOVING A VEHICLE ONTO A TRACK

5-14. Align the vehicle with the track and position a plank-type ramp on the end of the track (figure 5-9). When a ramp is not available, dig a shallow ditch where the end of the track can lie.

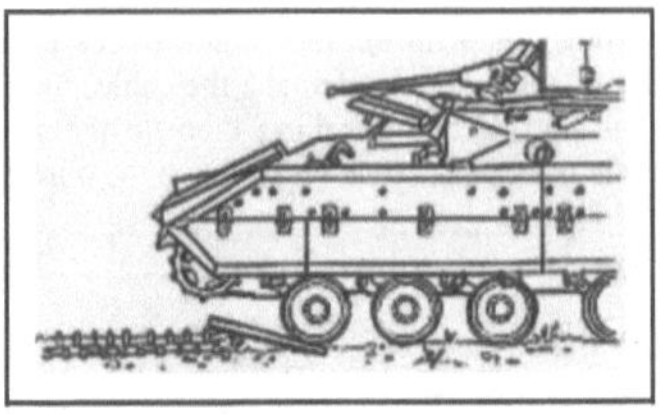

Figure 5-9. Moving a vehicle onto a track

INSTALLING A TRACK

5-15. To install a track—

- Align the track with the road wheels so that the center guide(s) will pass between the road wheels when the vehicle is moved (figure 5-10).
- Stop the vehicle when the rear road wheel is resting forward far enough for the entire track to pass over the sprocket.

Figure 5-10. Installing a track

- Tie a rope to the center of the track pin on the rear track link.
- Pass the rope over the center guide groove of the sprocket hub, around and between the rear support roller wheels, and back around the sprocket hub, making two turns.

5-16. As power is applied to the sprocket, and the free end of the rope is held taut, the end of the track is pulled up to the sprocket. Once the sprocket has engaged a minimum of three track links—

- Stop the sprocket, lock the brakes, and shut off the vehicle engine.
- Remove the rope from the sprocket hub and extend it forward over the compensating idler wheel.
- Restart the vehicle and move forward.
- When the end of the track has passed over the compensating idler, connect the track.

Chapter 6

Battle Damage Assessment and Repair Procedures

BDAR procedures apply to most operational levels (from the crew through field level maintenance) and depend on the extent of the damage, time allowances, available personnel with required skills, and accessible parts, tools, and materials. Maintenance personnel must act quickly to restore the vehicle to the combat-ready condition required to continue the mission or allow the vehicle to self-recover.

BDAR PRINCIPLES

6-1. To be effective, BDAR should follow certain basic guiding principles:

- Ensure standard maintenance practice is always the first consideration.
- Base decisions of using BDAR versus standard maintenance on the METT-TC.
- Provide an accurate assessment.
- Ensure economy of maintenance effort (use maintenance personnel only when necessary).
- Train multifunctional skills.
- Repair only what is necessary to regain combat capability.
- Remain flexible about repair priorities.

6-2. Commanders should address using BDAR in the logistics section of their operation order (OPORD). This will provide the crews and maintainers with a clear understanding of when and at what risk level they can perform BDAR. In wartime, BDAR may have to be liberally applied at the discretion of the commander. In military operations other than war, local command policy will direct the degree of BDAR to apply and when to use standard maintenance. However, commanders at all levels must ensure that both crews and maintainers perform annual BDAR training.

BDA ACCURACY

6-3. BDA is the necessary first step and starts the moment a crewmember recognizes battle damage. The assessment determines the extent of damage to include the subsystems and components damaged, the level of repair, and the risk involved. The assessment will also include an estimate of personnel, time, and materials required to perform expedient repairs. If not handled correctly, time, man-hours, parts, and resources will be wasted and opportunities to get back in the fight will be missed.

ECONOMY OF MAINTENANCE EFFORT

6-4. Maintenance assets will be heavily taxed on the battlefield. Because resources are limited (personnel, tools, and parts), it is imperative that maintenance resources are not wasted. Crewmembers must do repairs within their capabilities immediately rather than requesting maintenance personnel to do simple mechanical tasks.

MULTIFUNCTIONAL SKILLS

6-5. Personnel shortages and battlefield casualties mandate that maintenance team members have some knowledge of other skills needed to achieve critical repairs. A lack of key repairmen must not deter a team from doing battle repairs. Whenever possible, on-the-job training or cross-training of personnel should be done.

NECESSARY REPAIRS

6-6. On the battlefield, the objective is to return the system into battle with enough combat capability to get the mission accomplished. Cosmetic repairs are not necessary and are a waste of time and resources. If a broken item does not affect the ability to shoot, move or communicate, and does not pose a serious safety concern, it should not be repaired until the equipment is returned to maintenance where standard repair procedures can be performed.

FLEXIBILITY

6-7. Priorities of repair; a shortage of personnel, parts, and time; a chaotic battlefield environment; a change of mission; the lack of a weapon systems crew; and a host of unforeseen circumstances demand a flexible approach to BDAR. If specific repair techniques are not in the manuals, ingenuity can be the key to successful BDAR.

BASIC RULES OF ASSESSMENT

THINK SAFETY FIRST

6-8. Personnel must be aware of live/loaded ammunition, damaged weapons or ammunition, fuel/oil spills, and electrical cables and wiring.

- Look for unexploded ordnance (UXO) in the area before performing the assessment. Do not move or otherwise disturb UXO or any ammunition that has been fixed, armed, ignited, or involved in an accident until qualified UXO personnel have rendered it safe.
- Check the area for chemical contamination, to include depleted uranium (DU) when appropriate. Use a radiac meter to determine if DU is present from damaged ammunition or from armor that contains DU.
- Avoid environmental contamination by spills of fuel and oil. All spills should be reported through the chain of command to the unit's logistical element, such as the battalion S-4.

BEWARE OF BOOBY TRAPS

6-9. Booby traps and improvised explosive devices present a unique challenge when recovering abandoned vehicles. If equipment was abandoned or was unsupervised by friendly forces, the possibility of booby traps exists. To ensure the safety of BDAR/recovery operations, inspect equipment for tampering before attempting repairs.

RECOGNIZE BATTLE DAMAGE INDICATORS

6-10. Battle damage indicators (BDI) from an operator or crewman perspective include smoke, fire, unusual odor, unusual mechanical noise, leaking fluids, fault warming lights and alarms, and loss of mobility or system function. Immediate recognition and attention by crewmembers is important because BDI may not be apparent once equipment stops functioning.

PERFORM AN ASSESSMENT

6-11. The senior man present decides when and if BDAR is performed during combat. This decision is based on METT-TC and the appropriate risk repair level.

6-12. Do not attempt to operate systems or subsystems until the crew has performed an assessment to prevent further damage to equipment or personnel. For example, if all circuit breakers are tripped, including the main circuit breaker, the assessment process should lead to determining the best method or sequence required to restore power to the vehicle. This could be done by initially resetting the main circuit breaker followed by resetting the remaining circuit breakers one at a time.

6-13. In the forward battle area, the crew must attempt to move the vehicle to a covered or concealed position to prevent additional damage. The best "rule of thumb" is to move the vehicle at least one terrain feature or one kilometer away from enemy contact.

6-14. If the vehicle is non-self-recoverable, use any like or heavier class vehicle to recover the vehicle or to conceal it. If this is not possible, turn the turret (if the vehicle is equipped) in the direction of engaging fire to limit damage and provide return fire capability.

6-15. To enable a systematic assessment, crews and maintenance personnel should use the 11 basic steps to BDA. The 11 basic steps in this manual include—

- Visually inspect interior and exterior for damaged parts and systems.
- Visually determine if vehicle main systems appear to be operable.
- Perform equipment self-test function–using a built-in test, built-in test equipment, and a function test.
- Assess system performance (exercise each system if engine can be safely started).
- Determine which subsystems are affected.
- Determine if crewmembers can repair the damage. (Are enough crewmembers with the required skills available, and does METT-TC allow repairs at the current location?)
- Estimate the repair time (by crew and by MST).
- Estimate the number and type of repair personnel needed and the associated risk. Ensure command approval to perform repairs.
- Determine what materials are required.
- Determine what the vehicle limitations will be after repairing using BDAR or standard repair.
- Determine the recovery status—self/like/dedicated.

6-16. Based on METT-TC, the CRT/ MST/MT uses the assessment check information to—

- Determine if the current risk repair level assigned is appropriate for the required repairs for the CRT/MST/MT.
- Conduct tests with maintenance test equipment, if required.
- Perform additional vehicle operational tests, if necessary.
- Determine what must be repaired or replaced.
- Estimate repair times for each repair task.
- Determine the sequence and priority of the repair action.
- Determine the repair location and, if other than on-site, arrange for recovery of the vehicle to the repair site.

6-17. The following safety checks are performed to identify any obvious hazards.

- Is there a round of ammunition in the gun tube?
- Is any ammunition in a critical state due to shock, fire, or physical damage?
- Have any combustibles such as fuel, hydraulic fluid, or oil accumulated?
- Does wiring appear to be safe? Could an arc occur to stored ammunition or leaking combustibles?
- Is the fire-extinguishing system operational? If not, station a crewmember in the vehicle—prepared either to use a handheld fire extinguisher or to operate the onboard fire extinguishing system manually. Station a second crewmember outside the vehicle with an additional fire extinguisher.
- For systems with built-in self-test procedures, has a functional/operator test been performed on those systems that appear undamaged?

BDA GUIDELINES

6-18. This section gives guidelines for BDA. Use these guidelines to rapidly assess battle-damaged equipment and systematically determine which subsystems are affected, to include the time, personnel, and materials required for repair.

6-19. These guidelines will also assist in performing "equipment triage." Equipment triage is the process used to decide the order that battle-damaged equipment will receive repairs. This determination is based on combat or combat support equipment, time, urgency, materials, and personnel required to do the required repairs.

6-20. Units can develop locally produced forms or checklists that best support authorized equipment and unit maintenance structure, and that list personnel authorized by the command to approve BDAR actions based on the BDA. Consider the following guidelines (figure 6-1) when assessing battle damage. These guidelines can be tailored to fit your specific vehicle.

System Assessment Summary
☐ Determine vehicle status— ▪ Can the vehicle shoot, move, and communicate? ▪ Can the vehicle be repaired to shoot, move, and communicate? ▪ Can the vehicle be self-recovered, towed, or transported? ☐ Check engine, transmission, fuel system, electrical system, wheels and suspension, hydraulic system, armor/ammunition storage, armament/fire control, and communications to see if they can be repaired or recovered and identify any limitations. ☐ Identify expendables, parts, and tools and National Stock Number (NSN) if applicable. ☐ Estimate the time and personnel needed.
Hull Damage Assessment and Repair
☐ Record applicable NSN for exchanged and cannibalized parts. ☐ Check engine system—for example, starter, oil tank, air induction system, air cleaner, oil filter, drain valve, accessory drive, shaft, low oil pressure. ☐ Check transmission and final drive systems for faults—for example, transmission will not shift, broken linkage, vehicle will not steer, final drive locked, transmission leaks, parking and service brake serviceability, and oil cooler. ☐ Check fuel systems—for example, fuel tanks, fuel lines, fuel filters, fuel pumps. ☐ Check electrical systems—for example, wiring harness, slip ring, batteries, circuit breakers, and power distribution box. ☐ Check track and suspension systems—for example, compensating idlers, track adjusting link, road wheel arms, road wheels, support rollers, sprockets, shock absorbers, torsion bars, and track assembly. ☐ Check hydraulic systems lines and fluids, driver controls, and instruments.
Communications Damage Assessment and Repair
☐ Record applicable NSN for exchanged and cannibalized parts. ☐ Check serviceability of intercommunications, receiver, transmitter, antennas, cables, and security.
Turret Damage Assessment and Repair
☐ Record applicable NSN for exchanged and cannibalized parts. ☐ Check electrical system—for example, turret power, slip ring, circuit breaker, and wiring harness. ☐ Check armament—for example, bore evacuator, gun tube, breech group, and main gun mount. ☐ Check fire control system—for example, commander control handle and weapon sight, gunner primary and auxiliary sight, range finder, crosswind sensor, wiring system, gunner control handle, stabilization system, manual traverse and elevation, and loader's panel. ☐ Check hydraulic system—for example, auxiliary hydraulic pump, hydraulic fluid, and hydraulic reservoir.

Figure 6-1. BDA guidelines

6-21. BDAR may enable the equipment to either self-recover or continue the mission. The BDA will provide the commander with necessary information to make efficient decisions concerning whether to continue the fight or recover the equipment to the appropriate maintenance location.

6-22. Always report battle damage as soon as possible. Ensure that the damage is reported according to the local SOP and this manual.

MAINTENANCE PROCEDURES AND BDAR

6-23. The operator/crew prepares and provides the initial damage assessment and reports to the vehicle commander. They describe inoperable conditions, to include CBRN conditions and circumstances. (When the inoperable equipment is subject to or in danger of hostile fire, another vehicle can be used to recover it to a secure location.) The operator/crew assesses the situation and determines which type of maintenance support is required.

6-24. If directed, the crew shall proceed to make any field expedient repairs possible. Usually these repairs will consist of restoring firepower, communications, and/or vehicle mobility within the limit of their skills and the availability of materials and tools. They must also consider repairing items to make the equipment self-recovery capable. If repairs are beyond crew capabilities, they request assistance per the unit's standing operating procedure (SOP).

6-25. The vehicle commander will report the results of the crew/operator damage assessment to the platoon leader. He will name the major known causes of the vehicle's immobility and/or lack of firepower and/or communication failures. If repairs by the crew are possible, the vehicle commander will report the appropriate risk repair level, a total estimated repair time, and a list of functions that may be restored.

6-26. The platoon leader will respond with directives and, if required, will call a CRT to the location of the damaged vehicle for assistance. If possible, he will provide sufficient information to enable the CRT to bring any required repair parts, special tools, or recovery assets to the site.

6-27. Maintenance personnel will assess the equipment to verify the operator's/crew's damage assessment for accuracy or reconsideration of repair methods. Based on the maintenance assessment, the decision will be made to either attempt an on-site repair or request recovery assets to move the vehicle to a UMCP. METT-TC will determine if on-site repair or evacuation is necessary.

6-28. The CRT and MST will perform BDAR, using the BDAR kits and any other field expedient material on hand. Because standard maintenance repairs usually offer the best repair, maintenance personnel will strive to perform standard repairs—if METT-TC permits.

6-29. Equipment that is damaged but mobile may be used to move disabled equipment. If recovery vehicles are not available (and as the tactical situation permits), like or heavier classed vehicles may be used to recover disabled equipment.

6-30. If all critical repairs can be done with the skills, tools, and equipment on hand, the CRT (assisted by the crew) will proceed with the on-site repair. If the damage exceeds the repair capability of the CRT and time is available for an MST on-site repair, the CRT will call the MST.

6-31. Sometimes there is not enough time available for an MST on-site repair. If the vehicle is reparable, the CRT will recover the vehicle to a designated UMCP.

6-32. If the vehicle is not reparable, the CRT will provide—

- Recovery to the UMCP for evacuation to the rear.
- On site cannibalization, if approved by the commander and coordinated with support maintenance.
- Other needed replacement parts.

6-33. If the vehicle is contaminated, the CRT will mark the vehicle with the appropriate contamination markers and arrange for recovery to a decontamination site.

RECORDING BDAR REPAIRS

6-34. Attach a DD Form 1577 (Unserviceable [Condemned] Tag-Material) to all vehicle components (figure 6-2) that are repaired using BDAR or other expedient techniques. In an emergency, it is not necessary to complete the tag. The purpose of marking the component is to enable mechanics to recognize BDAR quickly when the equipment is subsequently returned for permanent repairs.

Figure 6-2. Example of a DD Form 1577

6-35. Because it is impractical to attach tags to some repairs located on the outside of the vehicle, place the tag in the compartment normally reserved for the equipment record book/folder or in an obvious place in the driver's compartment.

6-36. Place a tag conspicuously in the vehicle (in the driver's or commander's position) when a BDAR procedure has resulted in a degraded operating capability. Mark this tag "BDAR," and note its specific limitations or cautions.

6-37. When a component is cannibalized from a reparable vehicle, attach a tag in the space created by the missing part and near the master power switch. This will readily alert maintenance personnel that the part has been removed.

6-38. When a vehicle is recovered/evacuated for "standard" or "normal" repairs, a DA Form 2404 (Equipment Inspection and Maintenance Worksheet) (figure 6-3) is used. Also, a Unit Level Logistics System (ULLS)-generated Form 5988-E, Equipment Inspection and Maintenance Worksheet (Automated) (figure 6-4) is used as the electronic version of the DA Form 2404. Add the notation "BDAR" in the parts requested/maintenance faults section.

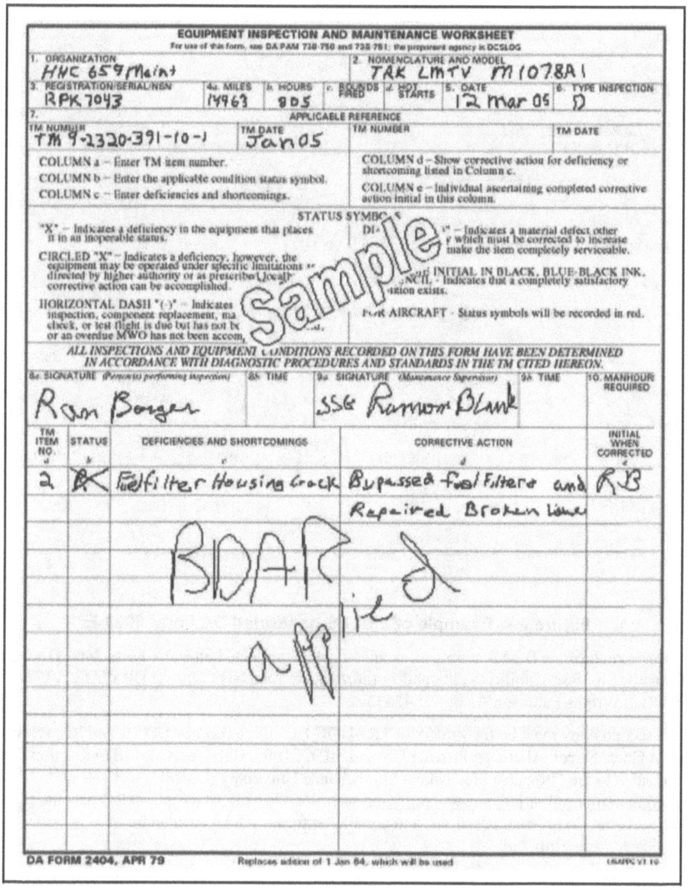

EQUIPMENT INSPECTION AND MAINTENANCE WORKSHEET

For use of this form, see DA PAM 738-750 and 738-751; the proponent agency is DCSLOG

1. ORGANIZATION	2. NOMENCLATURE AND MODEL
HHC 659 Maint	TRK LMTV M1078A1

3. REGISTRATION/SERIAL/NSN	4a. MILES	b. HOURS	c. ROUNDS FIRED	d. HOT STARTS	5. DATE	6. TYPE INSPECTION
RPK7043	14963	805			12 mar 05	D

7. APPLICABLE REFERENCE

TM NUMBER	TM DATE	TM NUMBER	TM DATE
TM 9-2320-391-10-1	Jan 05		

COLUMN a – Enter TM item number.
COLUMN b – Enter the applicable condition status symbol.
COLUMN c – Enter deficiencies and shortcomings.
COLUMN d – Show corrective action for deficiency or shortcoming listed in Column c.
COLUMN e – Individual ascertaining completed corrective action initial in this column.

STATUS SYMBOLS

"X" – Indicates a deficiency in the equipment that places it in an inoperable status.

CIRCLED "X" – Indicates a deficiency, however, the equipment may be operated under specific limitations as directed by higher authority or as prescribed locally, until corrective action can be accomplished.

HORIZONTAL DASH "(-)" – Indicates [illegible] inspection, component replacement, ma[illegible] check, or test flight is due but has not be[illegible] or an overdue MWO has not been accom[illegible].

DI[illegible] – Indicates a material defect other [illegible] which must be corrected to increase [illegible] make the item completely serviceable.

[illegible] INITIAL IN BLACK, BLUE-BLACK INK, [illegible]NCIL – Indicates that a completely satisfactory [illegible]ation exists.

FOR AIRCRAFT – Status symbols will be recorded in red.

ALL INSPECTIONS AND EQUIPMENT CONDITIONS RECORDED ON THIS FORM HAVE BEEN DETERMINED IN ACCORDANCE WITH DIAGNOSTIC PROCEDURES AND STANDARDS IN THE TM CITED HEREON.

8a. SIGNATURE (Person(s) performing inspection)	8b. TIME	9a. SIGNATURE (Maintenance Supervisor)	9b. TIME	10. MANHOURS REQUIRED
Ram Borger		SSG Ramon Blank		

TM ITEM NO. a	STATUS b	DEFICIENCIES AND SHORTCOMINGS c	CORRECTIVE ACTION d	INITIAL WHEN CORRECTED e
2	X	Fuel filter Housing Crack	Bypassed fuel Filter and Repaired Broken Line	RB

BDAR applied

DA FORM 2404, APR 79 — Replaces edition of 1 Jan 64, which will be used

Figure 6-3. Example of a DA Form 2404

DATE 20-APR- EQUIPMENT MAINTENANCE AND INSPECTION WORKSHEET FORM 5988-E

WK4WRC Unit

EQUIPMENT DATA

ADMIN NUM:	812	EQUIP SERIAL NUM:	050493
EQUIP MODEL:	M998	REGISTRATION NUM:	NG381A
EQUIP NOUN:	TRK UTL CGO 1.25T 4X4	INSPECTION:	W
EQUIP NSN:	2920811077153	READING:	M 010587

	NUMBER	CHANGE NUMBER
PUBLICATION	TM 9-2320-280-10	02
PUBLICATION	TM 9 [illegible]	0

SIGNATGURE: John Doe SIGNATURE: Ima Clerk TIME:______

Sample

PARTS REQUESTED

FAULT	DOC	NUM	NIIN	QTY DUE REC	STATUS DATE	DATE COMP	PRI	DLC
0001	3116	0001	000785961	00002 ___	___	0	13	Y
0002	3116	0002	000006881	00001 ___		0	13	Y

MAINTENANCE FAULTS

ITEM NUM	FAULT DATE	FAULT STATUS	FAULT DESCRIPTION	CORRECTIVE ACTION	INITIALS
0001	25-APR-	X	Won't Start	BDAR Battery Cable Repaired	J D W
0002	26-APR-	X	Fanbelt Broken 27 Apr 28 Apr	BDAR Linkbelt Installed	J D W
10	30 Apr	X	Engine Idles at 800 RPM	Adjusted Idle	J D W

BDAR

Figure 6-4. Example of a ULLS-generated DA Form 5988-E

6-39. When reporting a BDAR action, all details must be annotated on a DA Form 2404/DA Form 5988-E and forwarded to: Survivability/Vulnerability Information Analysis Center (SURVIAC, ATTN: AF-FDL-FES-CDIC), Wright-Patterson AFB, OH 45433.

6-40. A second copy must be forwarded to: TRADOC Executive Agency BDAR and Recovery, 5046 Havre de Grace Street, Aberdeen Proving Ground, MD, 21005. Comments and BDAR action forms can also uploaded to the Ordnance Mechanical Maintenance University (OMMU) web site at https://ommu.army.mil or http://ommu.army.mil web site. Users must have an AKO account to access the OMMU website. Comments, questions, and completed forms can be emailed to bdarfeedback@apg.army.mil.

6-41. These documents will serve as examples that can be tested to prove principle and may be included in future publications of this FM. Ensure all information provided is as accurate and detailed as possible.

> Note. The procedures in paragraphs 6-39 through 6.41 do not apply to USMC. USMC observes chain of command reporting procedures.

SPECIAL OPERATING ENVIRONMENTS

6-42. BDAR techniques may be more difficult in certain environments, such as extremely hot or cold climates. For example, certain molecular compounds (polymers) may take longer to cure in a cold environment but may cure very rapidly in hot environments. The instructions for these compounds are provided in the BDAR kits.

6-43. The BDA will be used to determine the time required to repair the system and whether repairs can be performed on site or if recovery is necessary. Whenever possible, standard maintenance repairs should be attempted. If that is not possible, then BDAR should be attempted to allow equipment to self-recover or continue the mission even though it may be at a reduced level of capability. Upon completing the mission and at the first practical opportunity, more complete, standard repairs should be performed to return the equipment to FULLY MISSION CAPABLE status. Whether to use cannibalization or controlled exchange is a decision rendered by the unit commander to whom the equipment belongs.

CANNIBALIZATION

6-44. Cannibalization is the authorized removal of serviceable and unserviceable parts, components, and assemblies from material authorized for disposal. Another possible source of cannibalization is captured or abandoned enemy equipment.

6-45. Supervised cannibalization and controlled exchange may be employed when parts are not available. Conditions for cannibalization are usually controlled by local policy and will be employed when addressed in an OPORD by higher headquarters.

CONTROLLED EXCHANGE

6-46. Controlled exchange is the removal of serviceable parts, components, and assemblies from unserviceable, economically reparable equipment. These parts are for immediate use to restore a similar item of equipment to a combat-capable or serviceable condition.

Note. Cannibalization and controlled exchange can be performed only when authorized by the commander.

BDAR IN JOINT OPERATIONS

6-47. Military units can expect to deploy as a component of a joint force. Maintenance personnel should work closely with other services to make collective use of tools and capabilities to perform BDAR. Despite differences in equipment and doctrine, the services have much in common that can be shared. Navy Seabees, Air Force maintenance activities, and most Navy ships have a machine shop and fabrication capabilities that would prove useful in supporting BDAR. Prior joint-service agreements should be further developed to use this capability between services. The same cooperation can be developed with allied nations. Most armies have a BDAR program. Many of the allied tools, materials, and techniques are similar to those of the United States. In addition, some foreign armies use our equipment, especially vehicles, and are a possible source for repair parts, but only if cannibalization is allowed by the HN. HN agreements should outline which BDAR or other maintenance services can be provided and the procedures required for obtaining support.

BDAR TOOLS AND EQUIPMENT

6-48. Special kits are available to support crew and maintainers to perform BDAR with access to BII and components of end items (COEI). These kits are allocated to the crew, CRTs, MTs, MSTs, and maintainers. Appendix A provides general information about the BDAR kits and the NSN for both crew and maintainer kits. BDAR kits allow repair in numerous areas—such as fuel, hydraulics, cooling, tires, electrical systems, and hull repair.

6-49. When possible, BDAR should first be performed by the crew using the crew/operator BDAR kit, BII, COEI, and AAL, OVE. Maintenance personnel will have access to the same items available to the crew/operator, as well as additional components.

HULL AND METAL COMPONENT REPAIR

6-50. It is critical to maintain hull integrity, especially during fording operations and when faced with a CBRN threat. Epoxies and polymers in the BDAR kit can be used to ensure hull integrity for small- to medium-sized holes (figure 6-5). Other components in the crew and maintainer BDAR kits—such as reinforcement tape and aluminum tape—can assist in patching armor tiles, vehicle fuel tanks, vehicle hulls, and any other metal surface or components on the vehicles. This capability allows military personnel to quickly reduce the effects of CBRN and other contaminants from entering their vehicle within minutes. Hull patches do not provide any additional ballistic protection; they are designed to maintain hull integrity (air/water tight).

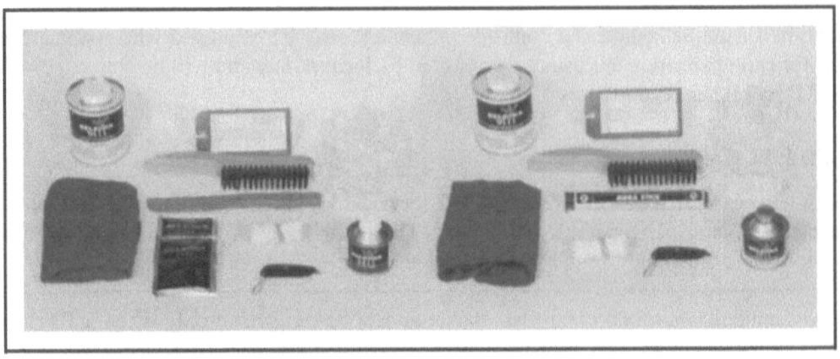

Figure 6-5. Hull and metal component repair materials

FLUID LINE COMPONENTS AND REPAIR

6-51. Fluid line repair items are in both kits. Hose clamps, duct tape, aluminum tape, and Belzona® SR Elastomer can be used to repair low-pressure rubber hoses (figure 6-6, left). Fittings needed for fluid line repair are found in the maintainer BDAR kit (figure 6-6, right). These fittings enable the maintainer to repair low- to high-pressure lines on most equipment.

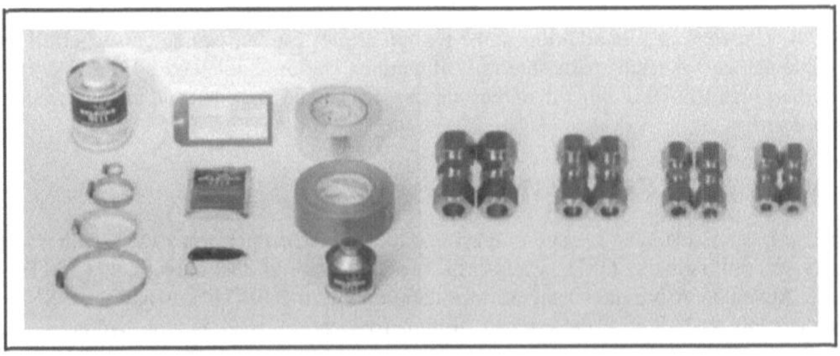

Figure 6-6. Fluid line repair items

LEAKING LOW-PRESSURE LINE FITTING

6-52. To repair a leaking low-pressure line fitting—

- Wind a string or rag tightly around the line behind the flare (figure 6-7). Wind the string clockwise in the same direction the coupling nut is turned for tightening.

- Slide the coupling nut over the material, screw it onto its connection, and tighten it securely against the packing string with a wrench. The string will act as a gasket and seal the leak.

6-53. Anti-seize tape is also available in the BDAR maintainer kit. Anti-seize tape prevents threads from seizing, but multiple wraps of this tape can serve as packing to seal damaged flared fittings and help seal threaded fittings. Polymers included in BDAR kits may also be used to repair leaking or damaged fittings because of their ability to flow into small spaces that caused the leak.

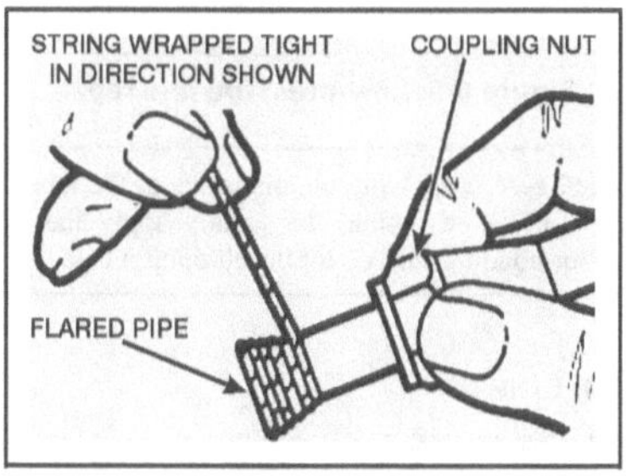

Figure 6-7. Low-pressure line fitting repair

CRACKED LOW-PRESSURE OIL OR FUEL LINES

6-54. Cracked low-pressure oil or fuel lines are usually caused by vibration or defective metal. If this occurs, stop the leak by wrapping the line tightly with friction tape held in place by wire. The wire helps the tape withstand pressure and usually stops the leak until a permanent repair can be performed.

6-55. Aluminum or electrical tape may also work if the surface can be cleaned to allow adhesion.

- Clean the surface of the line and cover just the crack with a couple wraps of rubber electrical tape.
- Follow up with multiple wraps of aluminum tape.
- Wrap wire around the tube or use hose clamps over the aluminum tape to reinforce the repair (figure 6-8).

6-56. Another quick method to repair a cracked line is with a small piece of hose.

- Use a piece of reinforced hose with an inside diameter equal to the outside diameter of the tube.
- Split the hose lengthwise.
- Coat the inside of the hose with sealant, if available.
- Install the hose over the leak with the split opposite the leak.
- Secure over the leaking area with hose clamps (figure 6-8). Additional clamps may be added to further reinforce the repair.

6-57. Polymers are also available in both BDAR kits to perform this repair. Polymers can be applied over the patch to provide additional reinforcement, or be included in the process before applying aluminum tape to ensure a better seal. Polymers do not require air to cure so the tape can be applied immediately after applying polymers.

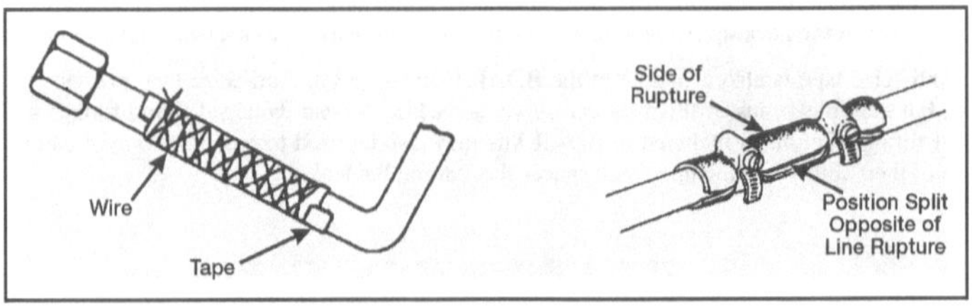

Figure 6-8. Low-pressure line repair

> ***Note.*** If polymer is used before applying aluminum tape, the tape must be applied before the polymer is allowed to fully cure. Failure to quickly apply the tape will result in irregular surfaces and create a poor bonding surface for the aluminum tape.

COLLAPSED FLEXIBLE FUEL LINE

6-58. To repair a collapsed flexible fuel line—

- Make a support for the inside of the fuel line by wrapping a thin wire around a pencil or another object slightly smaller than the flexible line inside diameter.
- Slide the coiled wire off the pencil and place it into the flexible line. Fuel will be able to flow and the fuel line will not collapse.

6-59. Applying external structure support may also repair a collapsed flexible line. If the line collapses during flow, but regains shape when flow stops, the following method may be used.

- Clean the exterior of the line with Elastomer conditioner included in the BDAR kit.
- Wrap the exterior of the line with wire, but allow at least 1/8 inch between successive wraps.
- Coat the entire affected area with Elastomer from the BDAR kit and allow it to cure. The Elastomer will bond to the flexible line and to the wire coil and provide external support to prevent collapse without restricting internal flow.

6-60. If the line does not regain shape when flow stops, remove the line from the closest source to the collapse and insert an object such as a pencil inside the hose to expand the collapsed hose. Then perform the repair listed in the above paragraph. Remove the object from the hose once Elastomer cures and reattach the hose to its original connection.

ELECTRICAL COMPONENTS AND REPAIR

6-61. Electrical repair capabilities are in both crew and maintainer BDAR kits. Electrical items and consumable materials can repair numerous electrical connections used in current equipment. Simple electrical repairs can be done with electrical tape, wire-nut connectors, 16-gauge or 22-gauge wire, and wire ties (figure 6-9). Pliers commonly found in most BII can act as cutters, crimpers, small bolt cutters, and wire strippers.

> **CAUTION**
>
> Vehicle power must be disconnected before working on any electrical wire.

6-62. Silicone sealant (included in the BDAR kits) should be applied to the inside of wire-nut connectors before installing them to seal the connection from moisture and corrosion.

LOOSE BATTERY CLAMP

6-63. When a battery cable clamp becomes loose and cannot be tightened—

- Remove the clamp; clean the post and the clamp as much as possible.
- Place the clamp on a flat surface and strike it with a hammer. This will compress the lead and close the hole on the inside of the clamp.
- Replace the clamp and tighten. If the clamp is still loose, use a nail or other metal object between the clamp and the post as a wedge to make contact between the battery post and the battery clamp.
- Loosen the clamp, insert the wedge between the battery posts, and tighten the clamp. If the clamp cannot be repaired, use a universal replacement battery clamp provided in the BDAR kit. The clamps in the kit are for positive battery posts, but they can also close tight enough to fit a negative post.

WARNING

When working around batteries and battery clamps, take care to prevent tools and jewelry from arcing. This could cause damage to vehicle electrical components and personnel injury. DO NOT smoke or permit an open flame near the batteries because gas from battery acid is explosive.

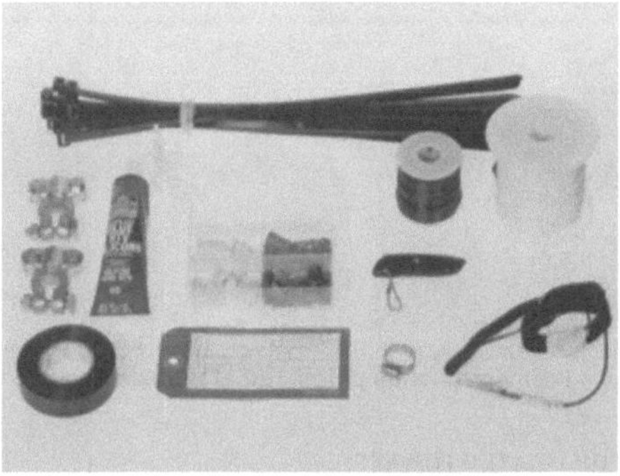

Figure 6-9. Electrical repair materials

TIRE REPAIR

6-64. There are two methods of tire repair. Tire repair capabilities (figure 6-10) are in both crew and maintainer BDAR kits. Fast, efficient tire repair is done with this kit if the hole is due to a nail or similar item causing small gashes. The tire plug kit is used in this case. If the tire damage is from shrapnel, the Belzona® SR Elastomer 2311 and reinforcement tape is used to patch the larger damage. Tire damage larger than 4 inches cannot be patched with this kit. Tire repairs can be made while mounted on the vehicle; however, the tire should not be under air pressure while patching.

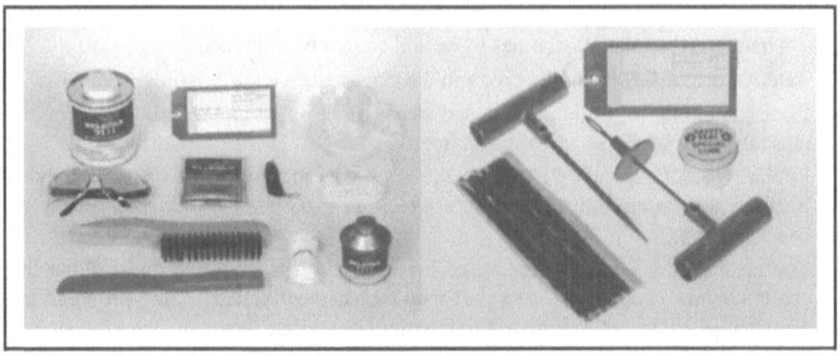

Figure 6-10. Tire patch materials and tire plug kit

COOLING SYSTEMS

6-65. Large holes in radiator reservoirs, fuel tanks, and oil reservoirs may be repaired using the environmental plugs shown in figure 6-11. These plugs serve to slow down leakage until a metal plug and other patching material are applied to the damaged area. The link belt (V-belt), included in the BDAR kit, is used as a replacement for fan, alternator, air compressor, and other belt-driven engine components. Radiator sealant can be used to seal small pinholes or seam leaks by pouring the sealant into the leaking radiator and replacing the radiator cap. Unit personnel are encouraged to add other items they feel are needed in their unit's BDAR kit.

Figure 6-11. Environmental plugs, radiator sealant, and a link belt

LEAKING RADIATOR OR HEATER HOSES

6-66. To repair a leaking radiator or heater hose—

- Allow the engine to cool so the hoses can be safely handled.
- Patch the leak by wrapping it with standard issue electrical or reinforcing tape.
- Refill the radiator with coolant and leave the radiator cap loosened.
- Operate the vehicle at a reduced speed until the hose can be replaced.

Polymers and radiator sealer are available in both BDAR kits to perform alternate methods of repair or reinforcement.

PUNCTURED TUBE-TYPE RADIATOR CORE

6-67. Radiators are often punctured when vehicles are operating in wooded or combat areas. When this occurs—

- Cut the cooling fins and push them away from the leaking tubes.
- Cut the leaking tube in half and fold the ends of the tube back approximately three-fourths of an inch (figure 6-12).
- Close the tube ends by pressing them flat with pliers.

6-68. Cooling system efficiency is reduced when several tubes are cut, causing the engine to overheat. When field expedient radiator repairs are made, loosen the radiator cap. This keeps the radiator pressure from building up and breaking the repair. High strength polymer kits are included in the crew and maintainer kit to cover the damaged core tubes after bending them over. Ensure the damaged area is very clean before applying polymer agents to ensure proper bonding. When these products are properly applied and cured, the radiator cap can be installed. Radiator sealer should be added after any major repairs are made to ensure small holes are sealed.

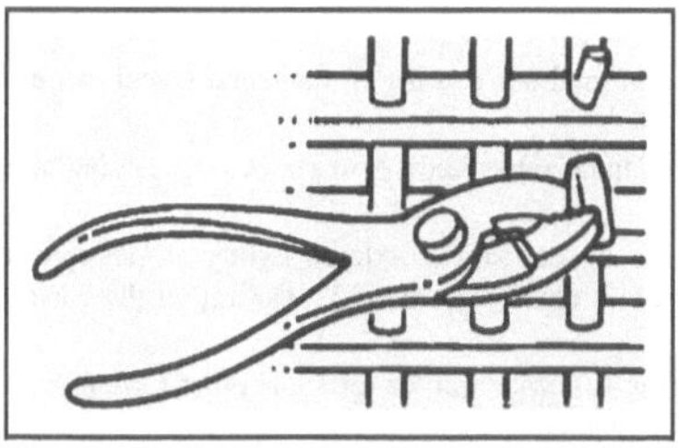

Figure 6-12. Tube-type radiator core repair

BROKEN FAN BELTS

6-69. When a fan belt breaks and a replacement is not available, mend or substitute it. To mend a broken fan belt—

- Punch holes in both ends of the broken belt and put a small gauge wire or bootlace through each hole.
- Secure the wire or laces.
- Replace the belt with just enough tension to drive the vehicle accessories.

6-70. To substitute for the broken fan belt—

- Use fiber rope from the vehicle tarpaulin or a piece of field telephone wire.
- Loop the wire or rope around the pulleys several times, pull as taut as possible, and tie with a square knot.

6-71. In both cases, operate the engine at low speed. This will prevent breakage or loss of the substitute belt. Both BDAR kits contain a link belt to temporarily replace a broken belt. Add or remove links from the belt to achieve the desired length.

BROKEN FAN BLADES

6-72. A broken fan blade will cause the engine to vibrate and make it dangerous to operate.

- On fans with an equal number of blades equally spaced, remove the remainder of the broken blade, and then remove the blade located opposite the removed broken blade. The vehicle can then operate, but be careful that it does not overheat.
- On fans with unequally spaced blades, the entire fan can operate under the light load for short periods of time. Trimming a small amount of fan blades opposite of the broken blade may also help to reduce vibration.

EXPEDIENT REPAIRS

DEFECTIVE TANDEM AXLE

6-73. A tandem axle with a burned-out bearing or damaged wheel can disable a vehicle or cause further damage if operation continues.

- Move the wheel of the disabled axle onto a rock, log, or similar object to raise the wheel as high as possible.
- While the wheel is raised, tie the axle as tightly as possible to the frame using heavy wire, ratchet strap, or a tow chain (figure 6-13). Do not let the chain, strap or wire cause damage to the brake lines.
- If the wheel bearing is burned out, or for some other reason the wheel does not turn, remove the axle shaft from the axle housing.
- Stuff the hole in the hub with rags to keep out foreign matter.

This expedient method allows the other wheels to drive. Reposition any cargo over good axles or transfer cargo to a second vehicle.

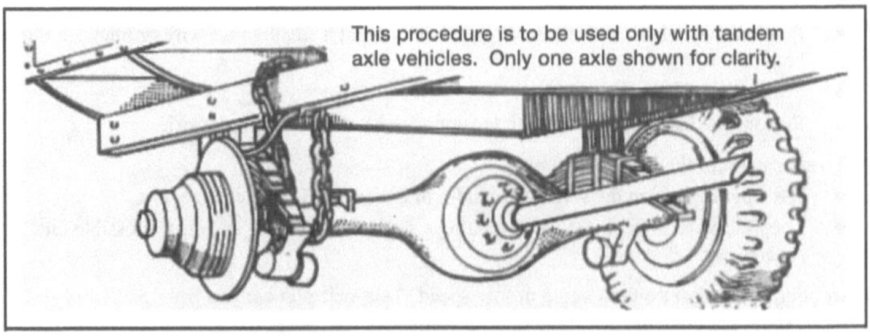

Figure 6-13. Attaching tandem axle to frame

6-74. Use the same technique on both ends of the axle if both wheels are defective. Since both ends of the axle are tied up, do not load the vehicle too heavily.

DEFECTIVE DIFFERENTIAL

6-75. If the defect is in the differential of a 4 by 4, 6 by 6, or 8 by 8 vehicle, remove the propeller shaft and drive axles (figure 6-14). For example, if the front rear differential is defective and the rear propeller shaft is removed, the vehicle can still be powered by the front wheels. When the axle shafts are removed, cover the openings securely to keep out dirt and foreign matter.

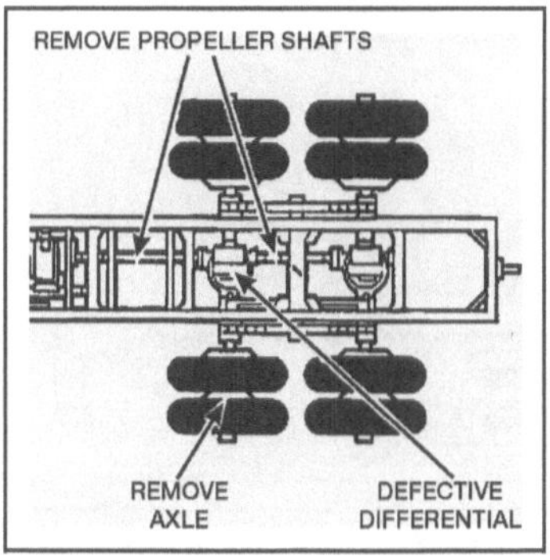

Figure 6-14. Defective differential

> **CAUTION**
>
> Avoid unnecessary oil spills by applying materiel from the BDAR kit over the holes. This also helps retain as much lubricating fluid as possible.

DAMAGED FRONT AXLE BRAKE SYSTEM

6-76. When damage has occurred to the front axle brake system, close the line from the junction block to the axle. If other damage to the air system exists, remove or repair leaking lines, install plugs where lines are removed, bypass brake components, or interconnect brake hoses to stop air loss. This helps the driver maintain steering control while braking and allows pressure to buildup to make rear brakes functional.

> **CAUTION**
>
> Report all corrective actions taken to stop air loss to maintenance personnel before disconnecting from the recovery vehicle.

DAMAGED ROAD WHEEL COMPONENTS

6-77. To repair a vehicle with a damaged road wheel, spindle, or road-wheel arm—

- Position the vehicle across a ditch.

Note. The ditch should be narrow enough to permit the front and rear road wheels to support the weight of the vehicle and deep enough to permit the track to sag away from the defective road wheel (figure 6-15). This will remove the tension from the torsion bar. If a ditch is not available, dig a trench.

- Tie the road-wheel arm up out of the way.

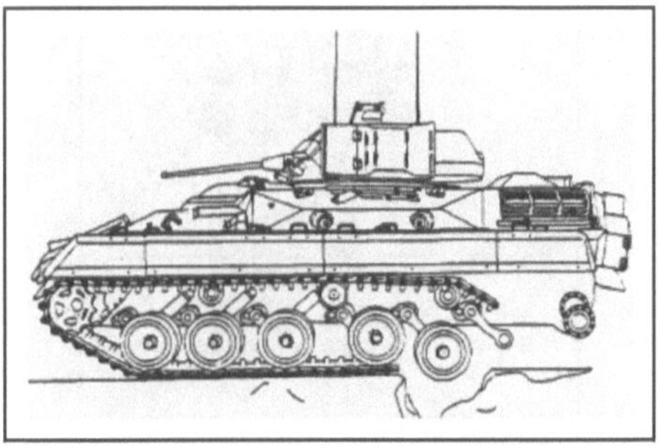

Figure 6-15. Suspending road wheel

- Remove the torsion bar by removing its cap screw and end plug, replacing the cap screw in the torsion bar, and prying behind the cap screw head with a tanker's crowbar.

- Reposition the vehicle on level ground.
- Remove the road wheel.
- Position a tanker's crowbar across two torsion bar support housings, and, using a rope from the vehicle tarpaulin, tie the road-wheel arm to the tanker's crowbar (figure 6-16).

Note. Use this expedient method with intermediate road wheels only.

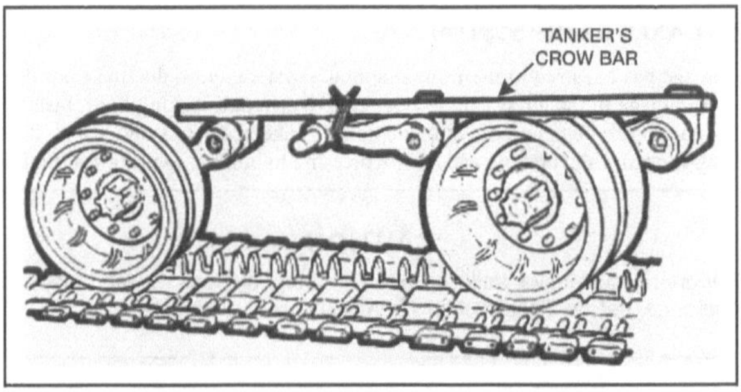

Figure 6-16. Road wheel expedient technique

Damaged Suspension Components

6-78. A tracked vehicle with a damaged track, rear road-wheel arm, or idler wheel can be operated, using the short track expedient method (figure 6-17). However, the hull must not interfere with the shortened track.

Note. This method is not recommended for AVLB or M1 family of vehicles.

6-79. To apply this expedient repair, remove the companion components to the damaged suspension parts—such as the rear shock absorber, track adjuster, road-wheel arm, and torsion bar. Before applying short track expedient methods, check the procedures outlined in the appropriate TM.

CAUTION

This procedure is time-consuming and may cause injury to personnel when performed.

6-80. The short track method is designed to regain mobility after component failure or combat damage. If the idler wheel and road arms number 5 and number 6 are damaged beyond repair, it is possible to short track to the number 4 road wheel. Whether damaged or not, the number 5 road wheel, road arm, and torsion bars must be removed to allow short tracking to the number 4 road wheel. If possible, reinstall the number 6 road wheel to improve vehicle stability. Sufficient track blocks must be removed to permit the track to be connected.

Note. If both sides of vehicle must be short tracked, remove all damaged road arms and idler wheels using the process described in paragraph 6-77.

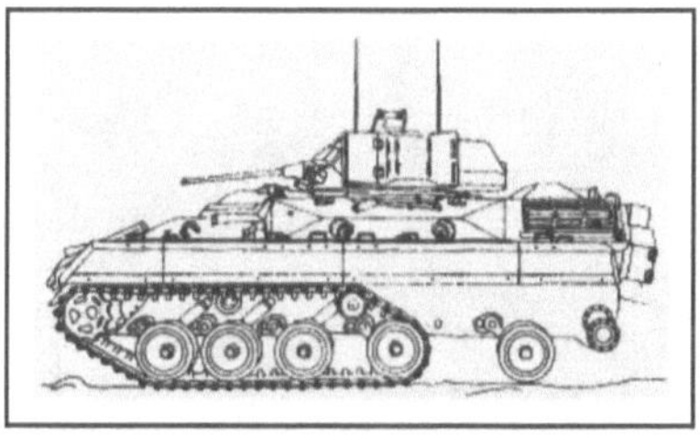

Figure 6-17. Short track expedient technique

SHEAR PIN SUBSTITUTE

6-81. Make a substitute for a broken shear pin:

- Punch out the remains of the broken shear pin, cut the remains in half, and insert the two shear pin halves with a short, wooden dowel between them (figure 6-18).
- Wrap friction tape around the shaft to cover the shear pinhole and prevent the end of the substitute shear pin from dropping out.

Note. Do not use a steel bolt, spike, nail, or screwdriver blade as a substitute shear pin because it could damage the winch and cable.

6-79. To apply this expedient repair, remove the companion components to the damaged suspension parts—such as the rear shock absorber, track adjuster, road-wheel arm, and torsion bar. Before applying short track expedient methods, check the procedures outlined in the appropriate TM.

CAUTION

This procedure is time-consuming and may cause injury to personnel when performed.

6-80. The short track method is designed to regain mobility after component failure or combat damage. If the idler wheel and road arms number 5 and number 6 are damaged beyond repair, it is possible to short track to the number 4 road wheel. Whether damaged or not, the number 5 road wheel, road arm, and torsion bars must be removed to allow short tracking to the number 4 road wheel. If possible, reinstall the number 6 road wheel to improve vehicle stability. Sufficient track blocks must be removed to permit the track to be connected.

Note. If both sides of vehicle must be short tracked, remove all damaged road arms and idler wheels using the process described in paragraph 6-77.

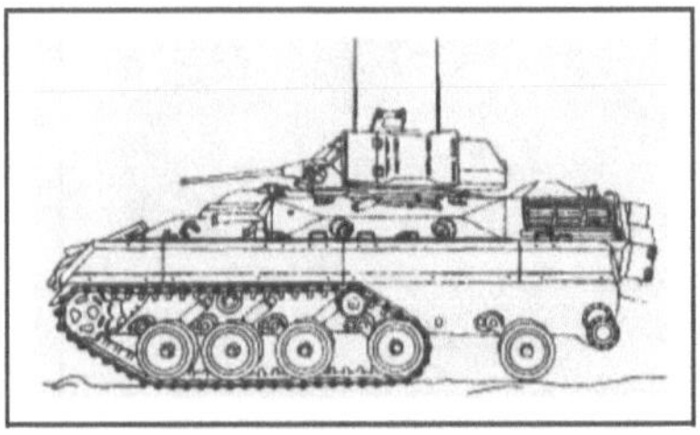

Figure 6-17. Short track expedient technique

Shear Pin Substitute

6-81. Make a substitute for a broken shear pin:

- Punch out the remains of the broken shear pin, cut the remains in half, and insert the two shear pin halves with a short, wooden dowel between them (figure 6-18).
- Wrap friction tape around the shaft to cover the shear pinhole and prevent the end of the substitute shear pin from dropping out.

Note. Do not use a steel bolt, spike, nail, or screwdriver blade as a substitute shear pin because it could damage the winch and cable.

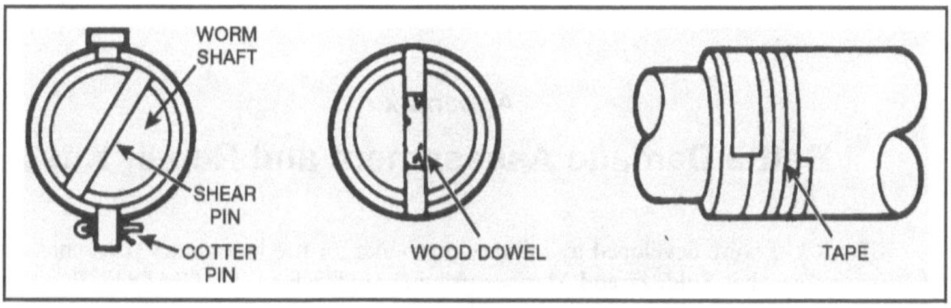

Figure 6-18. Shear pin substitute

Appendix A

Battle Damage Assessment and Repair Kits

BDAR kits were developed by military personnel for use by military personnel. It is imperative that Soldiers and Marines provide feedback to the TRADOC Executive Agency to upgrade the capabilities of the BDAR kits.

BDAR kits consist of essential material and capability items that enable efficient repair of common types of damage and random failures that occur during a mission. Each unit is expected and encouraged to modify its BDAR kit to fit specific operational needs and the unit's geographical environment. This appendix discusses the two primary types of BDAR kits.

Note. Additional information regarding BDAR kits and training can be found on the Ordnance Mechanical Maintenance University Web site at https://ommu.army.mil. Visitors are required to have an AKO username and password to access portions of this Web site.

CREW/OPERATOR BDAR KIT

A-1. The crew/operator BDAR kit (figure A-1) enables the crew to perform timely repairs. The recommended basis of issue (BOI) is one kit per crewed vehicle—for example, M1 Tank, M2 Bradley, M113 Personnel Carrier, M109 Howitzer, M9 ACE, M10 Stryker, and so forth. The crew/operator BDAR kit can be requisitioned with NSN 5180-01-502-9504.

MAINTAINER'S BDAR KIT

A-2. The maintainer's BDAR kit (figure A-2) supports operating forces. The recommended BOI is one kit per recovery, contact maintenance truck, and forward repair system—for example, M88, M936, M984, and M113. The maintainer's BDAR kit can be requisitioned with NSN 5180-01-502-9507.

Note. Each BDAR kit contains a current and complete component listing with NSNs to allow for inventory stock and replenishment.

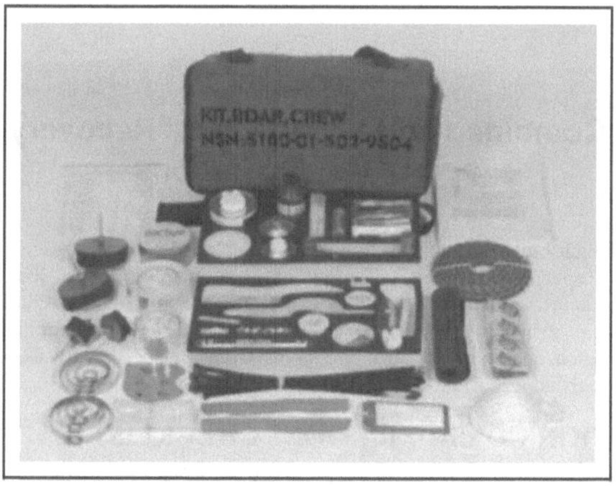

Figure A-1. Crew/operator BDAR kit

Figure A-2. Maintainer's BDAR kit

Appendix B

Combined and Multinational Recovery and BDAR Operations

Today's military missions increasingly call for combined and multinational recovery and BDAR operations to be part of a multinational force. For BDAR managers and operators, there are many opportunities and requirements to recover combined and multinational vehicles. This appendix provides guidance for coordinating and executing such operations. It also briefly discusses BDAR operations concerning captured or abandoned enemy equipment.

COORDINATION CONSIDERATIONS CHECKLIST

B-1. When participating in an operation in which U.S. assets may be used to support combined and multinational assets or vice versa, check the existing standardization agreement (STANAG) and SOP. Make contact with the affected combined and multinational unit to exchange information. Although coordination at initial phases of a combined and multinational operation will start at the highest level, as the relationship matures, coordination or information exchanges should routinely occur at tactical unit levels. This should be encouraged until continuous information exchange happens at the lowest level possible. The questions listed will become mission detractors if not clearly resolved before initiating BDAR missions. The following critical information should be exchanged, understood, and established during combined and multinational operations.

- Clearly establish command and control. Does a U.S. element revert to combined and multinational command and control for the duration of support to that combined and multinational unit, or does the U.S. parent organization retain command and control?
- Identify who establishes priorities for BDAR assets in an area where more than one command exists.
- Determine where recovery assets should be towed and by whom. Potential combined and multinational supporting units need the U.S. maintenance collection point (MCP) locations. U.S. forces need the MCP locations or other collection points established according to the supported combined and multinational unit.
- Identify the point of contact (POC) for questions and guidance. One POC is established for combined and multinational forces, and one for U.S. forces. U.S. elements need to contact these POCs.
- Establish the extent to which BDAR can be applied to combined and multinational units.
- Identify specifics regarding the primary vehicles that each nation might recover for the other.
- Exchange technical information regarding towing, preferred hookup locations for winching or overturned vehicles, and any other information that would assist in avoiding unsafe or dangerous BDAR operations.
- Exchange information regarding special actions required to secure sensitive items, such as radios, maps, signal operating instructions, or high cost or scarce components.
- Determine what the combined and multinational unit doctrine is concerning the use of the disabled crew on-site. Combined and multinational doctrine may be different from U.S. doctrine, which requires crews to assist in BDAR operations, as well as provide local security.
- Ascertain which type of coordination will be required concerning the passage of lines, if required. Clearly established POCs must be contacted for such passages.

- Exchange operational plans and graphics to preclude inadvertent distracters to combat operations or placing U.S. assets in unnecessary danger.
- Clearly establish recognition signals. These signals include challenges and passwords, as well as identifying vehicle markings. Recognition markings are especially important in operations where combined and multinational units and enemy forces use the same type of vehicle or in the case where the enemy may be using U.S. vehicles.
- Be aware of any special operational hazards, such as the use of CBRN elements or minefields. As necessary and where possible, arrange for combined and multinational guides or provide guides to U.S. supporting elements.
- If possible, provide combined and multinational units with U.S. BDAR kits for effecting BDAR on U.S. vehicles.
- If time and situation permit, arrange for mutual training or orientation sessions with counterpart personnel.
- If translations are critical for ongoing BDAR operations, arrange to have translators available. A better arrangement would be to have a technical advisor available from the nation owning the equipment.

EXECUTION CONSIDERATIONS CHECKLIST

B-2. The primary consideration is returning equipment to battle as quickly as possible while creating as little collateral damage as possible. Equally important is surviving to complete the mission. The following considerations involve approaching the site; local security, camouflage, and actions taken on contact:

- Before beginning recovery or BDAR operations on combined and multinational vehicles, ensure authorization has been given and obtain any necessary guidance.
- Attempt to locate a member of the crew or a technical representative to provide technical guidance.
- Before starting BDAR operations, obtain applicable manuals to determine proper BDAR actions. Even where language is a problem, pictures and diagrams may prove useful.
- Do not begin any operation until technical information has been obtained. Acting too quickly or prematurely might cause damage.
- Report completion of the mission to the U.S. chain of command. The U.S. chain of command will pass that information to the command and control at the liaison officer level.

SECURITY OF SENSITIVE ITEMS AND SALVAGE OF DAMAGED EQUIPMENT

B-3. Only divisional or higher commanders have the authority to order the destruction of equipment. With OPORDs, this authority is usually delegated to subordinate commanders. When a piece of equipment is destroyed, it must be reported through proper command channels.

SAFETY CONSIDERATIONS

B-4. Hazards that exist on the battlefield will also be present during the demolition of equipment (for example, toxic fumes and spilled fluids). Safety is an important consideration. BDAR personnel must become completely familiar with all aspects of the equipment being used. Applicable equipment technical manuals provide necessary warnings, cautions, and hazards.

B-5. Classified documents, notes, and instructions of any kind are removed from the vehicle before demolition. Classified materials must be rendered completely useless to the enemy.

Appendix C
Hand and Arm Signals

Visual signals are any means of communication that require sight and can be used to transmit prearranged messages rapidly over short distances. This includes the devices and means used for recovery operations.

VOICE CONTROL

C-1. Ground guides controlling all tracked vehicle recovery operations will use electronic voice means whenever available, supplemented by minimal hand and arm signals as the primary means of ground control during recovery and lift operations. Ground guides must also be familiar with recovery operations during hours of darkness—using a flashlight to augment hand and arm signals. Until a wireless system is developed, units will use clear voice capture (CVC) cables to link the ground guide with the vehicle operator via the vehicle intercom system for operations within 30 feet of the recovery vehicle.

C-2. An alternative means, especially for operations in excess of 30 feet of the recovery vehicle, is to connect a TA-312 (using an optional headset for hands-free operation) to the control box via WD-1 field telephone wire. If voice means cannot be established, hand and arm signals will continue to be used.

C-3. Restrictions for using hand and arm signals are as follows:

- Units must acquire extended CVC cables and/or other items needed in BDAR operations.
- Units are responsible for conducting familiarization training—stressing potential hazards for an extended cable or WD-1 wire to become snagged or severed during operations. The ground guides must be careful that the cable or wire does not become wrapped or entangled while moving. Therefore, if movement of the components is required, the wire or cable should be disconnected during such movements and reconnected when the ground guides are positioned safely.
- The crew must be informed that if voice cannot be established or fails at any time, they will return to hand and arm signals.
- The CVC cable assembly can be connected to any C-2298 box in a vehicle, except for the driver's box.
- When using either CVC cable or WD-1 wire with winching operations, the length must be such that the ground guide can be located safely and sufficiently outside any hazard area as required.
- Voice communications between the operator and the ground guide will make for safer operations by removing the doubt associated with hand and arm signals. These communications are particularly safer and more effective for limited visibility and night operations. They also remove doubt as to who is controlling the operator.

HAND AND ARM SIGNALS

C-4. All hand and arm signals are outlined in FM 21-60 and FM 21-305, which will be incorporated into the next release of FM 4-01.40. The most common types of visual signals are arm-and-hand, flag, pyrotechnic, and ground-to-air signals. Soldiers are not limited to the types of signals discussed and may use whatever mean is available. Chemical light sticks, flashlights, and other items can be used, provided their use is standardized within a unit and understood by all Soldiers and units working in the area. Some common arm-and-hand signals that might be used for BDAR operations are shown in figures C-1 through C-7.

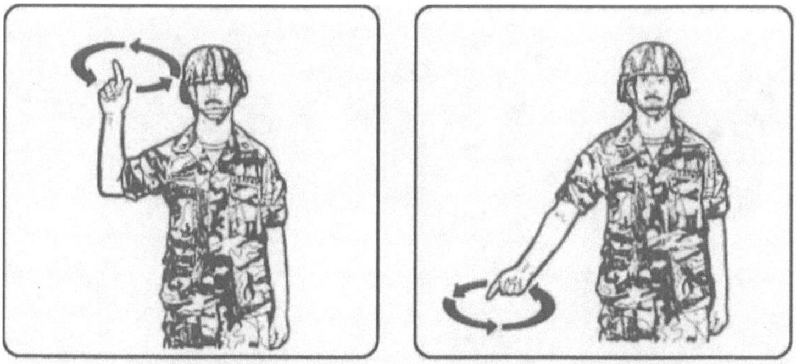

The signal to RAISE THE HOIST CABLE is made with the arm extended to the side and bent upward at the elbow, with the index finger extended from a fist, and the hand rotated slowly.

The signal to LOWER THE HOIST CABLE is made with the arm held downward and out slightly from the side, with the index finger extended from a fist, and the hand rotated slightly. This signal may be made with either hand with rotation made in either direction.

Figure C-1. Signals for raise the hoist cable and lower the hoist cable

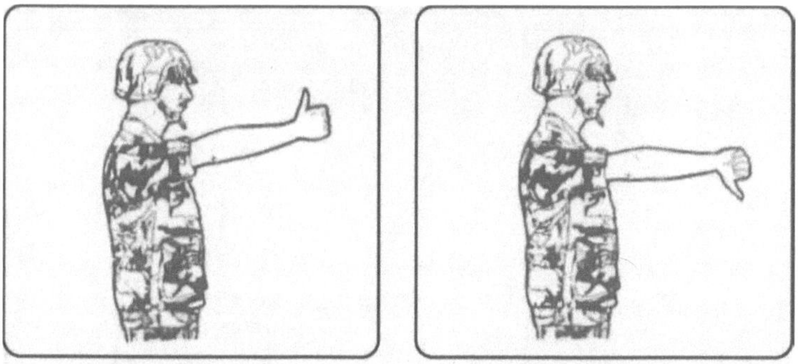

The signal to RAISE THE BOOM is made by extending the arm toward the operator with the fist clenched and the thumb pointing upward.

The signal to LOWER THE BOOM is made by extending the arm and clenching the fist in the same way, but with the thumb pointing downward. This signal may be made with either hand.

Figure C-2. Signals for raise the boom and lower the boom

The signal to RAISE THE SPADE is given by first pointing at the spade with the index finger of the left hand. While pointing with the left hand, extend the right arm toward the operator, with the arm bent upward at the elbow while clenching the fist and pointing the thumb upward.

The signal to LOWER THE SPADE is made in the same way, but with the thumb of the right hand pointing downward.

Figure C-3. Signals for raise the spade and lower the spade

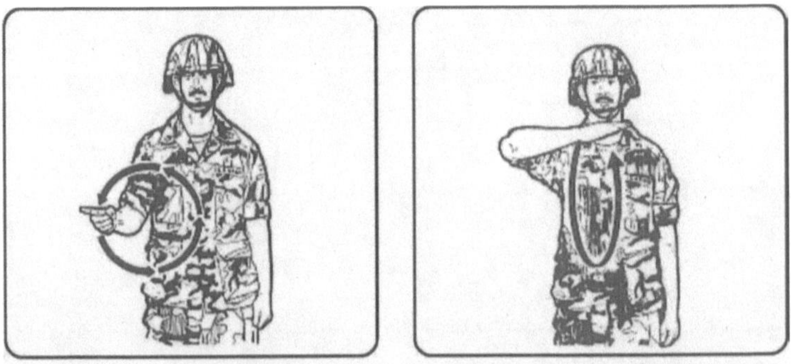

The signal to INHAUL THE MAIN WINCH CABLE is made by pointing at the operator with the index finger and rotating the arm in a circular motion.

The signal to PAY OUT THE WINCH CABLE is made with the arm bent, bringing the hand in front of the chest. The hand is moved down and away from the body at belt level, circling back to the chest. The circular motion is continued until the signal to stop is given.

Figure C-4. Signals for inhaul the main winch cable and pay out the winch cable

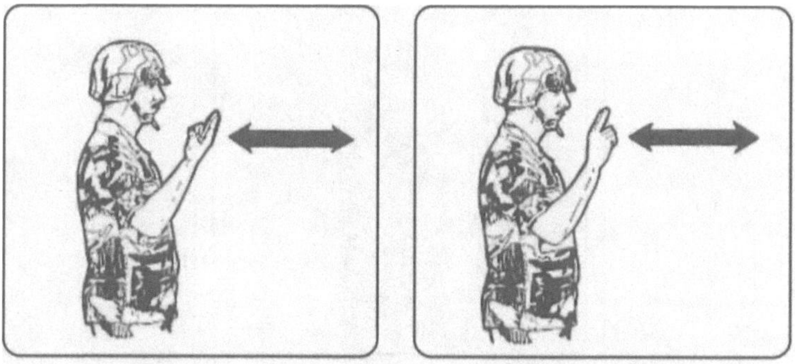

The signal to EXTEND THE BOOM is made with the index and center fingers extended upward, with the back of the hand facing the operator and the hand moved in toward and away from the chest by bending the elbow slowly in a pumping action.

The signal to RETRACT THE BOOM is made in the same way, but with the palm of the hand facing the operator.

Figure C-5. Signals for extend the boom and retract the boom

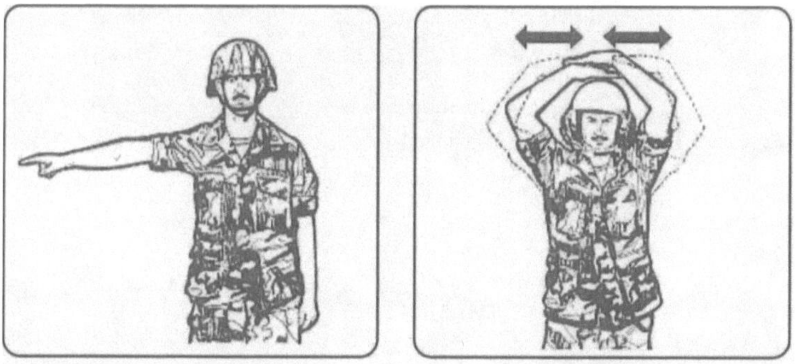

For vehicles with a traversing boom, the signal to SWING THE BOOM RIGHT OR LEFT is made by extending the arm to shoulder level in the direction that the operator must traverse.

For the signal to BUTTON UP, place both hands on top of the helmet, palms down, one on top of the other. Place both arms back and in the same plane as the body. For the signal to UNBUTTON, give the button up signal; then separate the hands moving them slighty to each side in a slicing motion. Repeat.

Figure C-6. Signals for swing the boom right or left and button up-unbutton

The signal to STOP any action that is being performed is given by clasping the hands together with palms facing each other at chin level.

Figure C-7. Signal for stop

Note. Refer to FM 21-60 for more information on hand and arm signals and using flashlights during night operations.

Appendix D

Recovery Guidelines for Operators/Leaders

Mission success on the battlefield may be linked to a unit's ability to perform vehicle recovery, to return immobilized equipment to operation, and to continue with the mission. Commanders must take aggressive actions to retrieve damaged equipment and return it to use. However, for a recovery operation to be successful, operators and leaders at all levels must be trained on recovery operations.

OPERATOR AND CREW

D-1. When the operator/crew detects disabled equipment, the damage is assessed and actions are initiated based on analysis results and the tactical situation. The crew/operator informs the chain of command of the status of the disabled equipment. Unit SOPs should prescribe notification procedures since these vary based on the type of unit, equipment, communications, and location of equipment.

D-2. The operator/crew will be trained to perform self-recovery and like-recovery on assigned equipment. This training will be practiced during garrison/field training exercises as prescribed in the unit's SOP.

D-3. The operator/crew normally remains with the disabled equipment to provide local security and until assistance arrives. When the maintenance personnel arrive, the operator/crew assists with the repair or recovery and stays with the vehicle until it reaches support maintenance.

Note. When the maintenance personnel arrive, the operator/crew assists with the repair or recovery and stays with the vehicle until it reaches support maintenance.

D-4. The following is a list of key items the operator/crew should know before requesting recovery from support elements.

- Location, map coordinates, and type of terrain.
- Nature of the disability.
- Tactical situation.
- Can BDAR be applied?
- Has BDAR been applied?
- Repair parts required, if known.
- Alternate radio frequencies.

RECOVERY PERSONNEL

D-5. Recovery equipment operators are usually highly trained mechanics and very familiar with the mechanical functions of equipment they must recover. These personnel must be skilled in the technical aspects of recovery, such as equipment rigging, towing, and uprighting procedures. They must also be skilled in related tasks—such as using the specialized BII on assigned equipment and operating in a tactical environment. Recovery equipment operators are assigned to company maintenance teams and to the recovery support section of the maintenance platoon. Personnel participating in recovery operations must be trained to check for and clear or disarm weapon systems of supported equipment. Specific procedures for the disposition of contaminated equipment, contingency plans, and any special tactical or security considerations should be covered in the unit SOP.

D-6. Recovery personnel are mechanics who perform repairs when not engaged in recovery missions. The following is a list of key items recovery personnel must know.

- Oxygen and acetylene tank operations for welding.
- Cutting torches.
- .50-caliber machine gun.
- Communications (both radio, and hand and arms signals).
- Map reading, compass use, and global positioning system.
- Chemical and biological agents.

D-7. Those conducting repair or recovery need to have a plan for recovery operations. The unit SOP will contain detailed checklists to assist in preparing for on-site support. Preparation should include—

- A verification of location and the status of disabled equipment.
- An update on the current tactical situation.
- A selection of primary and alternate routes.
- The availability of communications, to include communications checks, applicable call signs, and primary and alternate frequencies.
- Individual clothing and equipment, with emphasis on CBRN equipment.
- A basic load of rations and ammunition to support a 24-hour continuous operation.
- A selection of appropriate support equipment, vehicles, and personnel required for the mission.

D-8. Recovery teams need to be aware of classified communications devices and components, and other classified materials. This will assist with maintaining proper security and reducing chances of compromise.

LEADERS

D-9. Platoon leaders and platoon sergeants have the responsibility for coordinating recovery assets and manpower requirements for disabled equipment. This effort is performed simultaneously with the mission, and if the recovery mission interferes with combat operations or in any way compromises security, it must be coordinated with the tactical commander.

D-10. Leaders should be trained on the same tactical procedures as recovery personnel so they can periodically check the rigging and equipment for proper hookups and adjustments. (Special attention must be given to weight and clearance limitations when using bridges or underpasses.) The following is a list of factors that leaders should determine before supervising or requesting recovery support.

- Equipment identification.
- Alternate radio frequencies.
- Location (map coordinates if possible).
- Alternate routes (when possible).
- The condition of the disabled vehicle.
- On site repair capability.
- Repair parts required.
- The organic recovery capability.
- Tactical situation and security requirements, and risk level.
- Cargo, road, and movement restrictions.

D-11. The recovery manager and leader must be alert to new situations and changing requirements. Planning and prior preparation are needed for continued effective recovery support.

D-12. Specific leader, mechanic and operator BDAR training should encompass the following:

- Risk assessment procedures; see FM 100-14.
- BDAR assessment procedures; see chapter 6.
- BDAR TM familiarization.
- BDAR kit familiarization; see appendix A.

- Group equipment.
- Suspension systems (short tracking); see chapter 5.
- Electrical systems (bypassing components, wire repair); see chapter 6.
- Cooling systems (radiator bypass and repair); see chapter 6.
- Fuel systems (patching holes, replacing or making lines sections); see chapter 6.
- Hydraulic/oil systems (repair high pressure lines, repair oil lines); see chapter 6.
- Tire and track repair; see chapters 5 and 6.

Glossary

SECTION I: ACRONYMS AND ABBREVIATIONS	
AAL	additional authorization list
ACE	armored combat earthmover
AE	available effort
AKERR	allied kinetic energy recovery rope
AR	Army regulation
AVLB	armored vehicle launched bridge
BDA	battle damage assessment
BDAR	battle damage assessment and repair
BDI	battle damage indicators
BII	basic issue items
BMO	battalion maintenance officer
BOI	basis of issue
CBRN	chemical, biological, radiological, and nuclear
COEI	components of end-items
CRT	combat repair team
CVC	clear voice capture
d	diameter
DA	Department of the Army
DD	Department of Defense
DRMO	Defense Reutilization and Marketing Office
DU	depleted uranium
E	effort
F	fahrenheit
FM	field manual
FMTV	family of medium tactical vehicles
FOV	family of vehicles
ft	feet
FWTD	fifth wheel towing device
GCWR	gross combination weight rating
HN	host nation
HNS	host-nation support
lb	pound
LR	load resistance
MA	mechanical advantage
MCP	maintenance collection point

MCRP	Marine Corps Reference Publication
MCWP	Marine Corps Warfighting Publication
METT-T	mission, enemy, terrain and weather, troops and support available–time available
METT-TC	mission, enemy, terrain and weather, troops and support available, time available, civil considerations
MOPP	mission-oriented protective posture
mph	miles per hour
MST	maintenance support team
MT	maintenance team
NSN	National Stock Number
OMMU	Ordnance Mechanical Maintenance University
OPORD	operation order
OVE	on-vehicle equipment
POC	point of contact
POL	petroleum, oil, and lubricants
QSTAG	Quadripartite Standardization Agreement
ROMO	range of military operations
S-4	supply officer
SOP	standing operating procedure
STANAG	standardization agreement
T	ton
TB	technical bulletin
TM	technical manual
TRADOC	U.S. Army Training and Doctrine Command
U.S.	United States
UI	unit of issue
ULLS	unit level logistics system
UMCP	unit maintenance collection point
USMC	United States Marine Corps
UXO	unexploded ordnance
WD-1	field wire, parallel pair

SECTION II: TERMS

Battle damage — All modes of equipment failures that occur during operations in wartime and peacetime, which prevents that equipment from completing its mission.

Battle damage assessment — The timely and accurate estimate of damage resulting from the application of military force, either lethal or nonlethal, against a predetermined objective. Battle damage assessment can be applied to the employment of all types of weapon systems (air, ground, naval, and special forces weapon systems) throughout the range of military operations. Battle damage assessment is primarily an intelligence responsibility with required inputs and coordination from the operators. Battle damage assessment is composed of physical damage assessment, functional damage assessment, and target system assessment. Also called BDA.

Battle damage indicators — A measurable phenomenon, either quantitative or qualitative, that can be used to indicate the damage/change of a target. (JP 2-01) Also called BDI.

Battle damage indicator concept — The battle damage indicator concept can be related closely to equipment damage from the perspective of a crewmember. Battle damage indicators from a crewman's perspective include smoke, fire, unusual odor, unusual mechanical noise, leaking fluids, faulty warning lights and alarms, loss of mobility or system function. Any indication of trouble must be noted immediately because it may not be apparent once equipment stops functioning.

Battle damage repair — Essential repair, which may be improvised, carried out rapidly in a battle environment in order to return damaged or disabled equipment to temporary service. Also called BDR.

Brigade support area — A designated area in which the battalion field trains and combat service support elements from the division support command provide logistical support for a brigade.

Cannibalization — The authorized removal of serviceable and unserviceable assemblies and serviceable repair parts from unserviceable, uneconomically reparable, or excess end items of equipment authorized for local disposal.

Combat capable — The equipment meets minimal functional combat capability requirements.

Combat emergency capable — The vehicle meets the needs for a specific tactical maneuver or firing mission; however, all systems (shoot-move-communicate) are not fully functional.

Combat repair team — A team that consists of field level mechanics that should be trained in battle damage assessment and repair procedures. Also called CRT.

Fully mission capable	The equipment can perform all its combat missions for which it was designed without endangering the life of the crew. To be fully mission capable, the equipment must be complete and fully operable with no faults listed in the "equipment is not ready/available if" column of the operator's preventive maintenance checks and services -10 manual.
High-risk repair	Battle damage assessment and repair that may lead to further damage to equipment and/or injury to personnel. EXAMPLE: Emergency operation of an engine with inadequate lubrication may explode causing injury to personnel and cause further equipment damage.
Low-risk repair	Battle damage assessment and repair that may lead to minor equipment failures but has no chance of injuring personnel. EXAMPLE: Placing a plug in the face of a tire—this is not likely to cause more damage to the equipment or injury to personnel.
Maintenance support team	The team that the combat repair team calls when equipment damages exceed combat repair teams repair capability. Normally, maintenance support teams have more capabilities than those found in combat repair teams or maintenance teams. Also called MST.
Medium-risk repair	Battle damage assessment and repair that may lead to further damage of equipment. EXAMPLE: Bypassing a starter relay will allow the equipment to start but may cause excessive heat.
Non-self-recoverable capable	The vehicle cannot move under its own power but does not meet the criteria for cannibalization.
Payed out	The act of letting out or feeding cable or rope from its source.
Self-recovery capable	The vehicle can move under its own power and is able to clear the battlefield under its own power.
Unit maintenance collection point	A unit maintenance collection point located behind the forward line of own troops. It is the focal point of the unit/battalion maintenance effort established by the battalion maintenance technician and is near the unit's combat train. At this spot, he organizes maintenance personnel, recovery vehicles, tool trucks, and POL assets to support the battle. Work at this location is designed to return critical weapon systems to the battle. Equipment not repaired or repairable at the unit maintenance collection point is evacuated to a maintenance location where it can be repaired within acceptable time limits. Also called UMCP.

References

SOURCES USED

The following lists the sources quoted or paraphrased in this publication.

ARMY PUBLICATIONS

> ***Note.*** The Army has a new numbering system for field manuals. If an FM has been published with the new number, the bibliography lists the new number followed by the old number in parentheses. If the FM has not yet been published with its new number, it is listed under the old number.

AR 200-2. *Environmental Effects of Army Actions.* 23 December 1988.

AR 25-52. *Authorized Abbreviations, Brevity Codes and Acronyms.* 4 January 2005.

AR 385-10. *The Army Safety Program.* 29 February 2000.

AR 750-1. *Army Materiel Maintenance Policy.* 15 July 2005.

DA Form 2028. *Recommended Changes to Publications and Blank Forms.*

DA Form 2404. *Equipment Inspection and Maintenance Worksheet.*

DA Form 2406. *Materiel Condition Status Report.*

DA Form 2407. *Maintenance Request.*

DA Form 5988-E. *Equipment Inspection Maintenance Worksheet (EGA).*

DA Pamphlet 310-17. *Joint Interest List of Technical Manuals (JIL).* 01 October 1994.

DA Pamphlet 750-8. *The Army Maintenance Management System (TAMMS) Users Manual.* 22 August 2005.

DD Form 1577. *Unserviceable (Condemned) Tag-Materiel.*

FM 1-02 (FM 101-5-1). *Operational Terms and Graphics.* 21 September 2004.

FM 3-3. *Chemical and Biological Contamination Avoidance.* 16 November 1992.

FM 3-5. *NBC Decontamination.* 28 July 2000 (will be revised as FM 3-11.5).

FM 3-7. *NBC Field Handbook.* 29 September 1994 (will be revised as FM 3-11.7).

FM 21-60. *Visual Signals.* 30 September 1987 (will be revised as FM 3-21.60).

FM 3-25.26 (FM 21-26). *Map Reading and Land Navigation.* 18 January 2005.

FM 5-170. *Engineer Reconnaissance.* 5 May 1998 (will be revised as FM 3-34.212).

FM 5-125. *Rigging Techniques, Procedures and Applications.* 3 October 1995 (will be revised as FM 3-34.311).

FM 3-100.4/MCRP 4-11B. *Environmental Considerations in Military Operations.* 15 June 2000 (will be revised as FM 4-04.4).

FM 55-30. *Army Motor Transport Units and Operations.* 27 June 1997 (will be revised as FM 4-01.40).

FM 4-30.3 (FM 9-43-1). *Maintenance Operations and Procedures.* 28 July 2004.

FM 100-14. *Risk Management.* 23 April 1998 (will be revised as FM 5-19).

FM 7-0 (FM 25-100). *Training the Force.* 22 October 2002.

FM 21-305. *Manual for the Wheeled Vehicle Driver.* 27 August 1993.

TB 43-0142. *Safety Inspection and Testing of Lifting Devices.* 28 February 1997.

TB 43-0209. *Color, Marking, and Camouflage Painting of Military Vehicles, Construction Equipment and Materials Handling Equipment.* 31 October 1990.

TB 43-0239. *Maintenance in the Desert.* 1 May 2004.

TB 385-101. *Safe Use of Cranes, Crane-Shovels, Draglines and Similar Equipment Near Electric Power Lines.* 2 December 1987.

TB 600-1. *Procedures for Selection, Training, Testing, and Qualifying Operators of Equipment/Systems Excluding Selected Watercraft and Aircraft, Managed/Support by US Army Troop Support and Aviation Materiel Readiness Command.* 25 July 1977.

TC 9-237. *Operator's Circular Welding Theory and Application.* 7 May 1993.

TC 24-20. *Tactical Wire and Cable Techniques.* 3 October 1988.

TC 43-35. *Recovery Training.* 3 October 1995.

TM 9-1025-211-10. *Operator's Manual (Crew) for Howitzer, Medium Towed 155-mm, M198 (NSN 1025-01-026-6648) (EIC: 3EL).* 14 January 1991.

TM 9-2350-200-BD-1. *Operator's, Unit, and Direct Support Maintenance for Tank, Combat, Full-Tracked: 105-mm Gun, M1 (NSN 2350-01-061-2445) Tank, Combat, Full-Tracked: 105-mm, Gun, IPM1 (2350-01-136-8738) and Tank, Combat, Full-Tracked: 120-mm Gun, M1A1 (2350-01-087-1095) General Abrams (Hull).* 11 September 1991.

TM 9-2350-238-10. *Operator's Manual for Recovery Vehicle, Full Tracked, Light, Armored, M578 (NSN 2350-00-439-6242) (EIC: 3LA).* 27 April 1990.

TM 9-2350-252-BD. *Battlefield Damage Assessment and Repair of Bradley Fighting Vehicles Fighting Vehicle, Infantry M2 (NSN 2350-01-048-5920) M2A1 (2350-01-179-1027) and Fighting Vehicle, Cavalry M3 (2350-01-049-2695) M3A1 (2350-01-179-1028).* 11 September 1990.

TM 9-2350-256-10. *Operator's Manual for Recovery Vehicle, Full Tracked, Medium, M88A1 (NSN 2350-00-122-6826).* 31 March 1998.

TM 9-2320-260-10. *Operator's Manual for Truck, 5-Ton, 6X6, M809 Series (Diesel): Truck, Cargo: M813, M813A1 and M814; Truck, Bolster, Logging; M815; Truck, Wrecker, Medium: M816; Truck, Dump: M817; Truck, Tractor: M818; Truck, Tractor, Wrecker: M819; Truck, Van, Expansible: M820, M820A1 and M820A2 and Truck, Stake, Bridge Transporting: M821.* 14 June 1985.

TM 9-2320-272-10. *Operator's Manual For Truck, 5-Ton, 6X6, M939, M939A1, and M939A2 Series Trucks (Diesel), Truck, Cargo: 5-TON, 6X6 Drop side, M923 (2320-01-050-2084) (EIC: BRY); M923A1 (2320-01-206-4087) (EIC: BSS); M923A2 (2320-01-230-0307) (EIC: BS7); M925 (2320-01-047-8769) (EIC: BRT); M925A1 (2320-01-206-4088) (EIC: BST); M925A2 (2320-01-230-0308) (EIC: BS8); 6X6, M936 (2320-01-047-8754) (EIC: BTF); M936A1 (2320-01-206-4078)(EIC: BS6); M936A2 (2320-01-230-0304) (EIC:BTT).* 15 August 1996.

TM 9-2350-276-BD. *Operator's, Organizational, Direct Support and General Support Maintenance Battlefield Damage Assessment and Repair for Combat Vehicles.* 10 February 1984.

TM 9-2350-284-BD. *Battlefield Damage Assessment and Repair for Bradley Fighting Vehicles M2A2 (NSN 2350-01-248-7619) M3A2 (2350-01-248-7620).* 24 January 1992.

TM 9-2350-292-10. *Operator's Manual for Recovery Vehicle, Heavy Full-Tracked M88A2 (NSN 2350-01-390-4683) (EIC: ACQ).* 1 January 2002.

TM 9-2320-307-10. *Operator's Manual for Model XM1117 Armored Security Vehicle (NSN 2320-01-437-6957).* 1 April 2003.

TM 9-2320-356-BD. *Operator's, Organizational, Direct Support and General Support Maintenance Battlefield Damage Assessment and Repair for Tactical Wheeled Vehicles.* 18 December 1987.

TM 9-2350-358-BD. *Operator's, Unit, and Direct Support Maintenance Battlefield Damage Assessment and Repair for Howitzer Medium Self-Propelled: 155 mm, M109A6 (NSN 2350-01-305-0028) (EIC: 3FC).* 25 March 1994.

TM 9-2320-387-10. *Operator's Manual for Truck, Utility: S250 Shelter Carrier, 4X4, M1113 (NSN 2320-01-412-0143) (EIC: B6B), Truck, Utility: UP-Armored Carrier, 4X4, M1114 (2320-01-413-3739) (EIC: B6C).* 17 October 1997.

TM 9-2610-200-14. Operator's, Unit, Direct Support, and General Support Maintenance Manual for Care, Maintenance Repair, and Inspection of Pneumatic Tires and Inner Tubes. 1 September 2000.

TM 9-4020-200-10. *Operator's Manual for Allied Kinetic Energy Recovery Rope (AKERR) Kit (NSN 4020-01-211-8382).* 31 January 1989.

TM 38-410. *Storage and Handling of Hazardous Materials.* 13 January 1999.

OTHER REFERENCES

GTA 05-08-002. *Environmental-Related Risk Assessment.* 1 October 1999.

GTA 05-08-003. *Hazardous-Material-Spill Procedures.* 1 April 1998.

GTA 05-08-017. *The Environment and Deployment: Tactical Risk and Spill Reaction Procedures.* 1 January 2003.

STANAG 2113. *Denial of a Unit's Military Equipment and Supplies to an Enemy.* 10 October 1995.

STANAG 2399. *Battle Recovery and Evacuation Operations.* 6 April 1992.

STANAG 2400. *Battlefield Vehicle Recovery User Handbook (AEP-17).* 16 July 1996.

STANAG 4101. *Towing Attachments.* 21 February 2000.

DOCUMENTS NEEDED

These documents must be available to the intended users of this publication.

None

READINGS RECOMMENDED

These readings contain relevant supplemental information.

None

Index

Made in United States
Troutdale, OR
11/03/2026

56844645R00066